Social Aspects of Sustainable Dryland Management

Social Aspects of Sustainable Dryland Management

Edited by
DANIEL STILES
UNEP, Nairobi

Published on behalf of the United Nations Environment Programme (UNEP)
by

JOHN WILEY & SONS

Chichester · New York · Brisbane · Toronto · Singapore

Published in 1995 by John Wiley & Sons Ltd,
Baffins Lane, Chichester,
West Sussex PO19 1UD, England
Telephone National Chichester (01243) 779777
International (+44) 1243 779777

Other Wiley Editorial Offices

John Wiley & Sons, Inc., 605 Third Avenue,
New York, NY 10158-0012, USA

Jacaranda Wiley Ltd, 33 Park Road, Milton,
Queensland 4064, Australia

John Wiley & Sons (Canada) Ltd, 22 Worcester Road,
Rexdale, Ontario M9W 1L1, Canada

John Wiley & Sons (SEA) Pte Ltd, 37 Jalan Pemimpin #05-04,
Block B, Union Industrial Building, Singapore 2057

British Library Cataloguing in Publication Data

A catalogue record for this book is available from the British Library.

ISBN 0-471-95633-3

Typeset in 10/12pt Sabon by Dorwyn Limited, Rowlands Castle, Hants
Printed and bound in Great Britain by Biddles Ltd, Guildford and King's Lynn

This book is printed on acid-free paper responsibly manufactured from sustainable forestation,
for which at least two trees are planted for each one used for paper production.

Contents

List of Contributors

Dr Ron D. Ayling
Senior Research Officer, Forestry, Environment & Natural Resources Division, International Development Research Centre, 250 Albert Street, PO Box 8500, Ottawa, Canada K1G 3H9

Dr Solon L. Barraclough
UNRISD, Palais des Nations, CH 1211 Geneva 10, Switzerland

Dr Roy Behnke
Co-ordinator, Pastoral Development Network, Overseas Development Institute, Regent's College, Inner Circle, Regent's Park, London NW1 4NS, UK

Birget Duden
Ethnologist, Elangata Wuas Ecosystem Management Programme, National Museums of Kenya, PO Box 40658, Nairobi, Kenya

Dr Yvette D. Evers
Department of Anthropology, University College, Gower Street, London WC1E 6BT, UK

Francis Gichuki
Overseas Development Institute, Regent's College, Inner Circle, Regent's Park, London NW1 4NS, UK

Prof. Mamadou Bara Guèye
Associate in Drylands Programme, International Institute for Environment and Development, 3 Endsleigh Street, London WC1H 0DD, UK

Dr Sabine Häusler
Associate Researcher, Institute of Social Studies, Postbus 29776, 2502 LT The Hague, The Netherlands

Dr Robert K. Hitchcock
Co-ordinator of African Studies, Department of Anthropology, University of Nebraska, Lincoln, NB 68588-0368, USA

Dr Forouz Jowkar
Institute for Development Anthropology, 99 Collier Street, Suite 302, PO Box 2207 Binghamton, NY 13902-2207, USA

Dr Ilse Köhler-Rollefson
Co-ordinator, League for Pastoral Peoples, Pragelato Str. 20, 64372 Ober-Ramstadt, Germany

Dr Michael John Mortimore
Independent Consultant Researcher, Cutters' Cottage, Glovers' Close, Milborne Port, Sherborne, Dorset DT9 5ER, UK

Enrique Inatoy
KUNA Asociaciòn Kunas Unidos por Napguana, PO Box 536, Ciudad de Panama, Panama

Dr Judy Elisabeth Pointing
Research Officer, Institute of Development Studies, University of Sussex, Brighton BN1 9RE, Sussex, UK

Dr Bhakthavatsalam Rajasekaran
CIESIN, 2250 Pierce Road, Saginaw, MI 48010, USA

Dr Daniel Stiles
UNEP, PO Box 30552, Nairobi, Kenya

Dr Mary Tiffen
Overseas Development Institute, Regent's College, Inner Circle, Regent's Park, London NW1 4NS, UK

Dr Jorge E. Uquillas
Consultant Social Scientist, LAT Environment Division, World Bank, 1818 H Street, NW, Washington, DC 20433, USA

Dr Michael Warren
Centre for Indigenous Knowledge for Agriculture and Rural Development (CIKARD), Iowa State University, Ames, IA 50011-1050, USA

Foreword

The title of the workshop which gave rise to this volume—'Listening to the People'—is an important aspect of the new paradigm of sustainable development in the post-UNCED period. It imbues the concept of development with a meaning which goes far beyond the conventional one. Listening to the people is about building development around the priorities, needs and objectives of the people it seeks to benefit. It is about empowering people to gain control over their lives through active participation in their own development. It is about recognizing the value of indigenous knowledge in sustainable development. It is about the centrality of people in development projects rather than technology.

People can no longer be considered as abstract entities, as mere 'target groups' in the development process. Yet it is common for the implementing agencies to view people as 'problems' and regard themselves as embodying the 'solutions'. Clearly, such paternalistic or technocratic attitudes have not worked in the past and are doomed to failure in the future.

Most of the chapters in this volume recognize this. In fact, the failure of various dryland management schemes has been attributed, in these chapters, to the adoption of a top-down approach to development, ignorance of local systems, short time horizons and use of complicated, difficult to maintain systems. It has also been realized that success obtained in the laboratory may have little relevance at the ground level. It has even been found difficult to replicate success obtained in one region in another region. And yet there are success stories we can build on.

How do we replicate these success stories? What enabling mechanisms can be designed to increase people's participation in the process of managing dryland schemes? How can socioeconomic information be integrated into our dryland management programmes to provide effective assistance at the ground level? How do we maintain an effective two-way information link between the national governments, international agencies and local communities? How do we ensure that the benefits of sustainable development of drylands reaches the marginalized, the politically invisible masses?

What we require, in essence, are practical suggestions on the manner in which changes can be brought about which are both incremental and self-enforcing: for the people to be enabled to participate actively in their own development and for the implementing agencies to shed their negative attitudes about the people they seek to serve and benefit.

Another issue in drylands management which will be discussed here is the neglect of women and the factors that render them vulnerable. The brunt of the consequences of environmental degradation are borne by women. Environmental degradation contributes significantly to their work load while simultaneously reducing their capacity to meet their provisioning functions for their families. The 'invisible' nature of women's contributions needs to be recognized and their perceptions and needs should also be considered.

All of us are agreed that 'desertification' is a complex phenomenon. There has even been dispute about its precise definition. Yet the effects of desertification are manifested socially. The hardships suffered by the millions who stay behind in a land gradually losing all its productivity and the millions of those who decide to leave their impoverished surroundings for an even more miserable existence in an urban setting are the social manifestations of this malaise.

We all need to step out boldly from the cocoons of our disciplines to confront the social aspects of desertification in all their complexity, and furnish cogent and creative solutions which can help local communities, governments and assistance agencies in their negotiations of the desertification convention and in their deliberations on the practical applications of the concept of sustainable development.

Elizabeth Dowdeswell
Executive Director, United Nations Environment Programme, Nairobi, Kenya

Preface

The workshop on 'Listening to the People: Social Aspects of Dryland Management', held in Nairobi 14–18 December 1993, was organized by the Desertification Control Programme Activity Centre of UNEP to develop a better understanding of 'community participation' and 'bottom-up development'. There is a great deal of discussion in environment and development circles about the importance of involving rural communities in activities aimed at stopping dryland degradation, promoting good land management and achieving sustainable development. Experience has shown, however, that these objectives are extremely difficult to achieve in practice. A primary goal of the workshop was for the participants, who represented governments, donor agencies, the United Nations system, non-governmental organizations (NGOs) and local communities, to make recommendations of what needs to be done to achieve sustainable development in the drylands. These recommendations, presented at the end of this book, were passed on to the Intergovernmental Negotiating Committee for a Convention to Combat Desertification and to the Commission on Sustainable Development for their use.

The papers and discussions of this workshop analysed the experiences of over three decades of attempts by governments, donor agencies and non-governmental organizations to promote economic development in the drylands of developing countries. A common theme throughout the workshop was the fact that if strategies for dealing with land degradation were to be meaningful, they must result in fundamental changes in the power relationships between the various actors at international, national and local levels. Effective communication channels are needed in order to enable communities in drought-prone areas to express their needs and development priorities. Such channels would enable communities to negotiate with representatives of government and with other groups whose interests and activities have an impact on their livelihood. Participants called for reforming land tenure rights on the basis of existing systems of ownership or use, and for providing security by guaranteeing access to land and resources.

The workshop noted that a prerequisite for the success of any intervention affecting a local community is that the planners recognize the institutions, systems of indigenous knowledge and management structures that already exist. Participants stressed the importance of enabling women to participate equally in the decision-making process, and to increase their capacity for production, management and earning their own incomes. In particular, governments and international organizations should support local communities' efforts to develop environmentally sustainable systems for marketing renewable natural products from dryland areas.

The main problem that persists in promoting the 'bottom-up' approach is the bottleneck that exists in the vertical communications between local communities and the powers that formulate policy, planning and financing that affect local people's ability to make land management decisions. Developing effective lines of communication remains a challenge for the future.

Daniel Stiles
Nairobi, Kenya

Acknowledgements

I would like to thank sincerely Mr Franklin Cardy, director of the Desertification Control Programme Activity Centre of UNEP, for his initiative in providing the intellectual and financial support that enabled the *Listening to the People* workshop to take place. I would also like to thank Ms Mary Wariuki, Doreen Maina and Enid Ngaira of UNEP for their help in the preparation of this book.

Part I

SOCIAL DIMENSIONS AND CONCEPTS OF DESERTIFICATION

Although there appears to be a growing consensus among all interested parties that the social side of the land degradation problem holds both the source of the causes and the solutions, there are still many outstanding issues that need to be resolved. Chief among these is the concept of 'desertification' itself. Like beauty, it seems to owe much to the eye of the beholder. Until some consensus can be reached about the ecological processes of land degradation, debates will continue about the severity and extent of the problem, and of the long-term consequences of human pressures on land. The relationship of desertification and land degradation is also being debated. There are very divergent opinions, which will be explored in this part.

The linkages between land degradation processes and social phenomena such as management practices, land tenure rules, settlement patterns, indigenous knowledge systems, gender relationships and so on are very complex. These issues and linkages will also be examined in this part.

1 An Overview of Desertification as Dryland Degradation

DANIEL STILES
Nairobi, Kenya

INTRODUCTION

Since Aubreville (1949) first coined the term 'desertification' in the context of the humid and sub-humid zones of West Africa, there has been considerable discussion about what the term really means. More than one hundred definitions have been recorded (Glantz and Orlovsky, 1983; Warren and Agnew, 1988; Odingo, 1990a), and opinions vary greatly on what the concept of the phenomenon should include.

Some definitions include both climatic and human causes, others restrict it to human-caused degradation; some restrict the term to the drylands, while others think it should apply to more humid areas as well. The question of 'irreversibility' has been included by some, with all the controversy that this term invokes. More recently, debate has begun about the validity of using vegetation degradation as an indicator of desertification. Some have gone so far as to claim that desertification is a myth, and that it is not even occurring.

Because of the lack of any consensus on what desertification means, in spite of an internationally negotiated and accepted definition made at the United Nations Conference on Environment and Development (UNCED) in 1992, the term was intentionally avoided both in the title of the workshop that gave rise to this publication and in the title of this book itself. The authors in this volume, however, not only believe that desertification, i.e. dryland degradation, is occurring but recognize that it is a very serious threat to the well-being of the one billion or so people living in the drylands. The approach here does not focus so much on stopping or reversing desertification *per se* but rather on ways to achieve good land and natural resource management for sustainable development.

Various studies and publications since the mid-1980s question various aspects of the concept and extent of dryland degradation, and these have had significant consequences in political and policy-making circles, particularly in the industrialized countries. One reason for the weak support given to the proposal for a Desertification Convention by the North is thought by some to

Social Aspects of Sustainable Dryland Management. Edited by Daniel Stiles

he well-publicized claims that the United Nations (UN) has exaggerated the extent of the desertification problem, and that it has misrepresented the concept for political reasons (Thomas and Middleton, 1994; Pearce, 1994a–c; Helldén, 1991; Olsson, 1993a; Warren and Agnew, 1988).

The critiques enumerated above are having and will continue to have important negative consequences for achieving a better understanding of the dryland degradation issue and, more seriously, for the people who live there, if they result in the view that dryland degradation is not an important global problem. It is therefore essential that the misunderstandings, and even misrepresentations, that have been made by the critics be answered. If desertification is not properly understood then many of the proposals and recommendations put forth in this volume may be invalid or have no purpose. More importantly, such a misunderstanding will hinder implementation of the recommendations for action contained in the Convention to Combat Desertification.

DEFINITION

A representative sample of definitions will be presented in order to highlight the main points of controversy concerning the concept. The 1977 UN Conference on Desertification (UNCOD) defined desertification as (United Nations, 1978: 7):

> Desertification is the diminution or destruction of the biological potential of the land, and can lead ultimately to desert-like conditions. It is an aspect of the widespread deterioration of ecosystems, and has diminished or destroyed the biological potential, i.e. plant and animal production, for multiple use purposes at a time when increased productivity is needed to support growing populations in quest of development.

Further discussion in the UNCOD report and associated conference documents (United Nations, 1977) made it clear that the UN viewed people as a main causative factor in dryland degradation, though the process was complex and varied, and that the term should be restricted to the drylands. No official definition of the 'drylands' was given, though a detailed discussion of the concept was presented, along with a map showing the geographical distribution of various values of the Budyko–Lettau dryness ratio (Hare, 1977) and the 1977 UNESCO *Map of the World Distribution of Arid Regions*.

The UN also stressed that desertification was not something that emerged from deserts, carried by the hot, dry winds. It could occur anywhere where land was overexploited, and it was generally not correct to envisage it as an advancing wall of sand dunes or desert frontier. Rather, it is usually '. . . far removed from any nebulous front line' and it '. . . is a more subtle and insidious process' than the advancing desert front (United Nations, 1978: 5).

FAO/UNEP (1983) offered a revised definition of desertification in the con-

Photograph 1.1. The popular view of desertification is of sand dunes moving out of the desert to engulf villages and good land. While this does occur, much more land is lost to degradation away from desert margins. (Photo: FAO)

text of their efforts to develop a methodology for assessing and mapping desertification:

> Desertification is defined as a comprehensive expression of economic and social processes as well as those natural or induced ones which destroy the equilibrium of soil, vegetation, air and water, in the areas subject to edaphic and/or climatic aridity. Continued deterioration leads to a decrease in, or destruction of, the biological potential of the land, deterioration of living conditions and an increase of desert landscapes.

This definition included aridity in it, but again did not define its boundaries. FAO maps of desertification have, however, always excluded areas with more than a 180 days' agricultural growing period. It also introduced the concept of ecological equilibrium, one that is now under re-evaluation in dryland grazing ecosystems (see Behnke, Chapter 8, this volume). FAO/UNEP also viewed desertification as a *process*, going through several stages before reaching the final irreversible one. The processes were both natural and human, but desertification could only be slowed or stopped by human actions.

Dregne (1983: 5), a long-standing expert in desertification, offered the following definition:

> Desertification is the impoverishment of terrestrial ecosystems under the impact of man. It is the process of deterioration in these ecosystems that can be measured by reduced productivity of desirable plants, undesirable alterations in the biomass and the diversity of the micro and macro fauna and flora, accelerated soil deterioration, and increased hazards for human occupancy.

Dregne's definition is one of the few that does not mention climatic factors in the causation. The definition is also not restricted to drylands, but Dregne states (1983: 5) that he goes along with the general view that it should be. By using such terms as 'desirable' and 'undesirable' he also introduces the concept of a socio-economic rather than purely biological assessment of land degradation. An undesirable alteration in biomass could be bush encroachment into rangelands, decreasing their economic value for grazing livestock. In purely biological terms, however, bush would have raised productivity.

Nelson (1988) strongly criticized the entire concept of desertification as one that was poorly characterized and as a term that 'obscures its true shape' because of the diversity of definitions. This did not prevent him from adding to that diversity by offering his own definition (1988: 2):

Photograph 1.2. Land degradation starts with inappropriate use of fertile land, as here in Ethiopia where ploughing is being done on a steep slope with no terracing to stop soil erosion. (Photo: UNEP/ICCE)

> Desertification is a process of sustained land (soil and vegetation) degradation in arid, semi-arid and dry sub-humid areas, caused at least partly by man. It reduces productive potential to an extent which can neither be readily reversed by removing the cause nor easily reclaimed without substantial investment.

Nelson goes along with FAO in defining the upper limit of drylands as having no more than a 180-day growing season and a maximum of 1200 mm average annual rainfall. He recognized that degradation was naturally reversible, and that fluctuations occurred, so he introduced the concept of relative irreversibility. This he arbitrarily defined as a 10-year natural recovery period of productive potential, or a substantial capital investment to effect rehabilitation. Presumably, if natural recovery would take more than 10 years, or if the investment was uneconomical, then one could say that desertification was occurring.

FAO (1993) has more recently added biodiversity to its definition:

> Desertification is the sum of geological, climatic, biological and human factors which lead to the degradation of the physical, chemical and biological potential of lands in arid and semi-arid areas, and endangers biodiversity and survival of human communities.

Warren and Agnew (1988: 3) do not offer a definition of desertification of their own, but say after a review of definitions: 'The definitions do not distinguish between desertification (conversion to a desert) and processes that diminish rather than eliminate productivity without necessarily producing deserts, namely land degradation.' They take 'desertification' in a very literal sense as a process that should lead to simulated deserts. Thus, they see including waterlogging from over-irrigation and bush encroachment or increasing unpalatable species in rangelands not as desertification but as degradation. They also argue that sparser vegetation, i.e. lower biological productivity, can sometimes be more nutritive and desirable for livestock than more biologically productive vegetation, thus productivity itself is not a valid indicator. They think that desertification must cause permanent degradation and that vegetation resiliency means that vegetation degradation must be very serious to be an indicator of desertification.

These, and other, criticisms that were made prompted the UN to reconsider the official definition of desertification. An Ad Hoc Consultative Meeting of experts convened in 1990 decided that there was no point in distinguishing desertification from land degradation in the drylands, as this only confused the whole problem. What was of primary concern was the fact that land was degrading and producing less food and commercial output, resulting in increased hardship, poverty and migration. Whether land actually ended up looking like a desert was immaterial and not of relevance to the socio-economic questions, and these technical squabbles were diverting attention from the real issues of concern. The final definition adopted by the Ad Hoc group was (UNEP, 1991: 1):

Desertification/land degradation, in the context of assessment, is land degradation in arid, semi-arid and dry sub-humid areas resulting from adverse human impact. Land in this concept includes soil and local water resources, land surface and vegetation and crops. Degradation implies reduction of resource potential by one or a combination of processes acting on the land. These processes include water erosion, wind erosion and sedimentation by those agents, long-term reduction in the amount or diversity of natural vegetation, where relevant, and salinization and sodication.

After debate, UNCED added climatic variations to human impact as contributing causes in the definition. The final UNCED definition, approved by *Agenda 21* participating governments, reads: desertification is land degradation in arid, semi-arid and dry sub-humid areas resulting from various factors, including climatic variations and human activities.

Drylands were defined by an adaptation of the Thornthwaite moisture index of the ratio of precipitation to potential evapotranspiration (UNEP, 1991: 9):

Hyper-arid	<0.05
Arid	0.05–0.20
Semi-arid	0.21–0.50
Dry sub-humid	0.51–0.65
Moist sub-humid	>0.65

According to this definition of drylands, the areas by continent expressed in millions of hectares is as shown in Table 1.1 (UNEP 1991: 11).

Hyper-arid areas are considered to be unproductive land, except in very small pockets of favourability, and are therefore not included in measurements of desertification. In spite of the fact that the international community formally accepted at UNCED the recommendation of a group of dryland experts to include vegetation as an important indicator of land degradation, some still do not accept the new definition and wish to distinguish desertification from land degradation as something unique (Thomas and Middleton, 1994).

Desertification is best seen as land degradation taking place in the drylands as defined above, following the same principles and processes as those seen in other eco-climatic zones. It is a cluster of processes which can fluctuate, with

Table 1.1.

Type of land	Africa	Asia	Australia	Europe	North America	South America
Hyper-arid	672	277	0	0	3	26
Arid	504	626	303	11	82	45
Semi-arid	514	693	309	105	419	265
Dry sub-humid	269	353	51	184	232	207
Total	1959	1949	663	300	736	543
%	32	32	11	5	12	8

Photograph 1.3. The eventual result of soil erosion is unusable land

periods of regeneration, and it is only irreversible economically in its mid to later stages. Its nature and causes will be particular to any given situation, depending on the natural ecosystem variables and history of land use. It is normally a very slow process, and thus can be assessed only over decades of observation, not years. It is rarely ecologically irreversible, though natural regeneration would only occur from a severe state either in the absence of human pressure or under exceptionally good management practices. It is nothing mysterious and singular, and the term is more of a political symbol than a scientific expression.

Thomas and Middleton (1994) have taken the views expressed by Warren and Agnew (1988) and made a book of them. These two works contain a number of incorrect statements of fact and misunderstandings (Stiles, 1994 and in press), which have led some journalists to write articles with titles such as 'Encroaching deserts "are a myth"', 'Treaty without a cause' (referring to the Desertification Convention), 'The myth of the marching desert', and 'Deserting dogma'. Even supposedly serious articles have used ostentatious titles such as 'Sandstorm in a teacup? Understanding desertification' (Thomas, 1993).

WHAT ARE THE MYTHS?

Although it is impossible to deal completely here with all of the theoretical and technical aspects of the critics of desertification, it is important to examine some of the more important statements that have been made, and the basis of these statements. The major criticisms and their proponents are as follows.

Lund University

A widely quoted figure of 'desert advance' of an average 5.5 km a year made in an unpublished report by Lamprey (1975) has been wrongly used by some as representing an official United Nations statement (Helldén, 1984, 1988, 1991; Olsson, 1985, 1993a,b; Thomas and Middleton, 1994). They have indiscriminately linked Lamprey's report to UN reports (UN, 1977; UNEP, 1984) and statements by others on the rate and extent of desertification, presenting them all as one package. In doing this, they have managed to accuse the UN, and UNEP in particular, of promoting a misleading image of desertification as the advancing desert, even though UNEP publications and reports explicitly have tried to dispel that notion.

Remote-sensing studies have been made by geographers from Lund University in Sweden, best represented by the Helldén and Olsson references cited above, that conclude that everything said in the Lamprey report was mistaken, and that there was no evidence of any long-lasting land degradation in the part of Kordofan Province of Sudan that was studied. The Lund University team further has stated that the UN desertification assessments (UN, 1977; UNEP, 1984, 1991) were made with inadequate data and that they have exaggerated and misrepresented the extent and nature of the desertification problem.

The real myth makers

Warren and Agnew (1988) and Thomas and Middleton (1994) have used the Lundites' early conclusions about the lack of evidence for land degradation, and have gone along with them in criticising UN statistics on the rate and extent of desertification, particularly the 1984 report. They have stated that inaccurate statistics have led to a false appreciation of the problem, which has resulted in inappropriate action. Thomas and Middleton (1994) and some journalists have also focused inordinately on the 'desert advance' image, creating a straw man to blow down. All these critics are disturbed by the lack of an agreed-upon and proper definition of the term 'desertification', and claim that the concept has been oversimplified and/or badly characterized by the UN.

The major criticisms can be reduced to:

(1) Lamprey's report, in particular the widely quoted 5.5 kilometres annual desert advance south.
(2) The UN statistics of the rate and extent of desertification.
(3) The definition and characterization of desertification.

Although the critics lump together the first two items above, the question of Lamprey's report and the UN assessments must be delinked, because they are very different in nature.

Examination of the Lund studies

Hellldén (1984) and Olsson (1985) used aerial photos from 1962 and Landsat multi-spectral scanner (MSS) satellite imagery from 1972/3 and 1979 of east-central Kordofan in Sudan to conclude that no consistent trend of a degrading landscape could be found, and that declining crop yields were due mainly to a lack of rainfall, not land degradation. These conclusions have since been expanded by Olsson (1993a,b) and others (Hellldén, 1991; Warren and Agnew, 1988; Thomas and Middleton, 1994) to suggest that desertification has much less to do with declining food production than previously thought.

Although it would seem rather obvious that rainfall would be more important for crop yields than any change in soils over a relatively short period, and also for the state of natural vegetation production than any measurable human-induced degradation, the methods and data that the Lund team present are surprisingly weak when examined closely. The following points question the validity of their studies:

(1) It was shown as early as 1980 that the MSS satellite imagery could not differentiate important vegetation classes, such as grassland, from bushland and woodland (Hellldén, 1980: 23) and that there were serious problems in identifying land degradation using MSS satellite imagery (Hellldén, 1980: 43). These problems persisted with Olsson, and he could not differentiate most vegetation types, including *bare soil* from grassland, bush-grassland and tree-grassland (1985: 65).
(2) To properly interpret satellite imagery, ground-truthing must be carried out to verify what the registered reflectance wavelengths are representing on the ground. It is best to carry out the ground-truthing as close as possible in time to the registration date of the imagery (Hellldén, 1980: 11). It is also crucial that the sample ground-truth locations be known with

geo-referenced accuracy in order to locate them on the satellite image. The 'ground-truthing' done by the Lund team satisfied none of these criteria.

Ground-truthing was carried out in 1980, 1982 and 1983 while the imagery was from 1972/3 and 1979. The truthing was done from a car following tracks, and positioning of samples was by dead-reckoning. No foot transects were described, and land classification was made subjectively by a visual assessment of the landscape. Most samples were taken at villages (Olsson, 1985: 21), thus the areal coverage was very limited and highly biased. Some quotes:

> There was . . . an average positioning error in the study area of 1 to 2 km. It was, for example, in no cases possible to determine the position [of a sample point] to a certain Landsat pixel (Olsson, 1985: 22–23).
>
> It has been impossible to include any field observations taken simultaneously with a satellite passage. This means that absolute calibrations of satellite data to measured ground conditions have not been carried out (Olsson, 1985: 32).
>
> The correlation between image data and ground truth data suffered most likely from the poor geometrical precision of the field data (Olsson, 1985: 68).
>
> [We] assumed that data collected through this procedure [the ground-truthing by car], probably resulting in biased data but still treated as a true random sample, were representative enough to describe strata units . . . The procedure, as it was applied in Kordofan, is probably *not accurate enough for significant change studies*' (my italics) (Helldén, 1984: 20).

The above quotes, and much more information contained in the reports, makes it clear that no valid ground-truthing was carried out. In other words, there was no way to know what the false colours in the satellite images represented, rendering any interpretations suspect. They had no way of differentiating bare ground and/or sand from grassland or savanna, so how could they find out if degradation was taking place?

They concluded that decreases in rainfall were more responsible for crop yield declines than soil degradation. This conclusion appears reasonable, but it is useful to look at their method of analysis. Olsson (1985: 113) could find no significant correlation between rainfall amount and crop yields, so he selected instead a variable that did show a good correlation: number of days with at least 1 mm of rain. He left out the drought years of 1968–74, however, as these would have lowered his correlation (Olsson, 1985: 113). Even with this fudging, Olsson (1985: 116) encountered other problems: 'The data set used may contain severe errors. Especially the data on agricultural productivity contain errors of unknown magnitude.'

The agricultural yield statistics are collected unsystematically and subjectively by inspectors, with no field measurements, thus any correlation between rainfall and yield per hectare would be highly suspect. It is also difficult to assess the results since no tables with rainfall or crop yield data were presented. Their figure 7.10, which presents frequency polygons of these variables, seems to show that there is no consistent correspondence between years of a high number of >1 mm rainfall days and millet yield, although it was stated that 71% of the variation in crop yields could be explained by rainfall parameters. What about the other 29%? They make no attempt to explain it.

Olsson (1985: 118) found extremely low levels of organic carbon in the soil, and lower levels of soil nutrients were correlated with higher agricultural intensity. He also stated that the low organic carbon, '. . . indicated a soil surface subject to erosion or exhaustion. Once this stage has been reached, further land degradation may not affect the amount of nutrients available for the crop . . . [I]t also indicates that continuing soil erosion and exhaustion in the cultivated area have very little effect, since the nutrient status has been almost totally depleted' (Olsson, 1985: 118). Both Helldén and Olsson noted that fallows had ceased to exist in many parts of the study area, with fields being continuously cultivated. What Olsson is saying in essence is that *the land is so desertified that further degradation cannot take place.*

Another serious criticism of their methodology is the fact that they each have only two points in time of satellite observations, one during a drought (1972/3) and one in a good rainfall year (1979). They strongly criticize others for using only two reference points in time: 'It is impossible to say anything about trends by comparing the conditions on two occasions only' (Olsson, 1985: 17), yet that is exactly what each of them did: 'In this study I have also compared two occasions only, but the first one during a period of drought and the second during more favourable climatic conditions' (Olsson, 1985: 147). The 1962 aerial photos they frequently mention in order to create the impression that they have a long time series are not even shown in any reports, and are inconsequential to their study conclusions.

They did not address many of the Lamprey observations that were evidence of desertification:

- Lamprey stated that his evidence for a desert margin move south was based on a vegetation shift of the *Acacia-Commiphora* zone and the *gizu* ephemeral grazing vegetation from their locations indicated on a 1958 map. Helldén (1984: 28, 1988: 9) and Olsson (1993b: 24) have repeatedly claimed that Lamprey used the 75 mm rainfall isohyet on the 1958 map to mark the desert boundary, when actually he was using vegetation types. From this mistaken assumption, the Lund team concluded that Lamprey had located the desert boundary 90–100 km further north in 1958 than it actually was. They apparently did not look for the vegetation types in the specific areas indicated by Lamprey.
- Lamprey noted that there was an extensive die-off of *Acacia senegal* woodlands along the 14th parallel, with replacement by the largely useless *Leptadenia* shrub. The Lund data showed an increase in *Leptadenia*, which they failed to highlight or connect with Lamprey's observations.

- Lamprey made several observations of new sand encroachment in specific areas. The Lund team concentrated on one mobile sand dune complex and ignored the rest. Even the one they looked at, in the southern Kheiran area, had moved. They concluded, however, that they could find no 'significant shift' of sand dunes, not defining what 'significant' meant to them. In most of the cases cited, however, Lamprey was not even referring to dunes, he was talking of sand sheets. Since these could not be detected by the satellite imagery, the Lund team did not address the question.

Thus, despite acknowledging the weaknesses of their own data, Olsson (1993a,b) and Helldén (1991, 1994) continue to make strong claims that there is no evidence for land degradation in Kordofan, and they have extended this claim to the entire Sahel. Helldén (1994) is currently trying to create the impression that land degradation cannot be demonstrated in the Sahel. However, a close reading of his conclusions includes the proviso 'by means of repeated satellite observations' (1991: 383). Degradation has been well attested by other means (UNSO, 1992), and satellite imagery has been used in studies in the Sahel and elsewhere to document land degradation (see Odingo, 1990b for reports on studies in Mali, Kenya and elsewhere; Ringrose and Matheson, 1986; Dalstead, 1988; Sehgal *et al.*, 1991). Why are the Lund team's results so different from those of everyone else?

There is much more that could be said about the Lamprey report and the Lund studies, but the above should suffice to demonstrate that the so-called scientific data accepted uncritically by Warren and Agnew (1988), Thomas and Middleton (1994) and others are of low scientific standards.

Desertification rate and extent

The UNEP (1984) report did not contain accurate statistics. The data simply do not exist. This is not to say, however, that UNEP exaggerated the scale of the problem. Warren and Agnew (1988: 5) claim that UNEP's figure of 35% of the earth's surface threatened by desertification includes arid non-productive areas not under threat, and from this they and others (Thomas and Middleton, 1994) say that UNEP is exaggerating the problem. In fact, the 35% of the earth's surface referred to does *not* include non-productive hyper-arid land, but only productive land (UNEP, 1984: 12; Mabbutt, 1984: 104). Warren and Agnew (1988: 35) further erroneously confuse the UNEP/FAO desertification *hazards* map with that of a *status* map, and conclude that there are numerous and obvious inconsistencies. Since they apparently did not know what they were looking at, inconsistencies were inevitable.

It is a point of considerable debate how vegetation should be viewed in assessing desertification (see Barraclough, Chapter 2, this volume). Thomas (1993) claims that since vegetation degradation should not be assessed as a desertification indicator, the problem is much smaller in extent than UNEP presents it. This question is a very complicated one, as it involves concepts of vegetation resiliency, regeneration, social versus biological standards of vegetation productivity, and so on.

The critics' view is perhaps too narrow. It overlooks the fact that biomass provides over 90% of household energy and construction in most dryland rural areas. Wild plants are also sources of medicines, food, ritual objects and raw materials for utensils (Stiles and Kassam, 1991). The view also ignores the question of undesirable species increase. The importance of vegetation for soil conservation and livestock production are well known, and livestock are major sources of food and wealth for dryland people. The fact that vegetation can regrow is no reason to ignore the grave socio-economic consequences of vegetation degradation.

There is an extremely important aspect of this question that has been underestimated, however—that of time depth and long-term rates of change.

Most researchers think that 30 years of observations is a long time, but paleo-environmental change is normally observable only over centuries or millennia (Stiles, 1981, 1988). The regeneration of land seen after a severe dry season or drought that seems to impress supporters of the 'new paradigm' so much should be examined very closely in the context of dozens of such cycles of dry and humid conditions over very long periods. People have been using the lands under consideration here for millennia; land degradation did not suddenly appear in the twentieth century.

Degradation occurs in small increments, and under good climatic and/or management conditions land can even regenerate to a better state than the immediately preceding one. But it might not be a state as good as a century ago. Degradation and these fluctuations should be viewed over centuries of time, and the long-term trend in drylands seems clear. What is currently lacking are data on natural vegetation and crop productivity and soil features from the time land first came under exploitation by humans, to use as a base line against which to assess current degradation status. To illustrate this concept, a caricature will be presented using oversimplified and hypothetical data:

> In the year AD 1700 an average hectare of uncultivated land in Kordofan was producing 2000 kg of dry matter and it contained 100 trees/shrubs. The soil organic carbon content was 10% and there was a humic top layer of 5 cm depth. The land was cleared for cultivation. In 1701 an average hectare produced 1000 kg of millet. Over the years there were losses and gains of soil nutrients, variations in crop production and wind erosion blew away soil at an average of 0.1 cm/yr over almost 300 years. Fallow periods decreased, however, and in 1995 the average hectare is under almost permanent cultivation, abandoned in drought years, but returned to when rainfall conditions improve. There is no topsoil layer at all, and the first 10 cm depth of soil contains 0.5% organic carbon, with none below that. Crop yields are highly variable from year to year, but the maximum

> since 1970 has been 500 kg/ha. When the land is abandoned, the natural vegetation production is 265 kg of dry matter per hectare. There are an average of 10 of the old tree/shrub species on each hectare, but many more *Leptodenia* and *Callotropus* (useless colonizers).

The above is the minimum of what would be needed to assess desertification properly. It would be better if data existed on demographics, settlement patterns, levels and methods of technology, management practices, etc. Archaeology and related studies can provide some of the needed information, but when lacking it entirely conclusions that land degradation has not taken place in a certain area should be made with caution.

What few people seem to appreciate is that the current and historically recent (i.e. the past 50 years) observations are not the complete picture when assessing land degradation. The zero point to start measuring change is not 30 or 50 years ago, it is the time from which people started using a piece of land. When viewed in this context, assessments of land resiliency, regeneration, dryland robustness and so on will become more balanced. This is not to deny that drylands are resilient and have considerable powers of regeneration, only that it should be recognized that xerophytic plants evolved their adaptive strategies to deal with highly variable soil moisture and grazing/browsing pressure prior to Neolithic times and the pressures exerted by humans and their domestic plants and animals.

Warren and Agnew (1988: 11) conclude that little was done up to 1984 to implement the UN Plan of Action to Combat Desertification (PACD) because the statistics were not believed by anyone. Such a conclusion is hard to justify, since the statistics and assessment appeared simultaneously in the General Assessment of Progress report (UNEP, 1984). Few paid much attention to the very gross statistics presented in 1977 at UNCOD. Subsequent to 1984, as the critics say themselves, many national and international environmental, scientific and lay publications repeated the UNEP findings, and they were used extensively by governments of affected countries to lobby for assistance. If the 1984 figures were not initially accepted, why were they so widely quoted and used?

Why more was not done had nothing to do with the UNEP statistics, it had to do with priorities and policy decisions made by donor and recipient governments (Stiles, 1984). There are a host of problems in dryland developing countries, and land degradation is only one of them. Donors tend to concentrate on projects that have attractive economic returns, which usually are located in agriculturally high potential or urban areas. There are understandable reasons why large investments are not made in areas that provide low economic returns and have sparse human populations. Accepting this conclusion means that inexpensive ways of tackling the land degradation problem in drylands must be found, and some of the chapters in this volume address that issue.

The data are getting better (UNEP, 1991, 1992), as even the critics recognize (Thomas and Middleton, 1994), but national programmes of desertification assessment and mapping need to be carried out. There is a long way to go before accurate statistics are available on the rates and extent of desertification in different parts of the world, but this should not influence an appreciation that the problem is great and growing.

Definition and characterization

Although some people do not accept the UN definition of desertification, a better understanding of and consensus about land degradation is approaching as a result of detailed research (e.g. Behnke *et al.*, 1993; Greenland and Szabolcs, 1994). The claim made by many that UNEP has characterized desertification as a marching desert is simply not true, as a review of their publications will show.

UNEP documents and publications describe desertification in terms remarkably similar to those of its critics. In fact, some UN documents are written by its critics (UNEP, 1992 was written by Thomas and Middleton; Andrew Warren was the main author of UNSO, 1992). Some of the critics, such as Warren and the Lund geographers, participate in UNEP expert group meetings and help to shape the policies and activities they criticize. It is critics such as Helldén and Olsson who dwell upon desert margin areas, and make generalizations about global dryland degradation based on samples from the Sahara fringe, which in terms of areal extent and population numbers is a very small part of the dryland degradation problem, though the problems of the people who live there should not be minimized.

It is ironic that at one time, even Helldén (1981: 7) was stating: 'Desert boundary oscillations in the southern Sahara and increasing growth of desert patches in Tunisia were indicated in two of the studies'. Olsson (1983) originally concluded that there was land degradation in Kordofan, caused by overcultivation, but he subsequently changed his mind.

CONCLUSIONS

The term 'desertification' is better viewed as a political statement than as a scientific expression. It has served its purpose in raising awareness about environmental problems and sustainability in the drylands, but there is little purpose in expending energy and resources debating the definition and concept independently from land degradation in general. There is still a way to go before land degradation is perceived and analysed in a proper time perspective.

The precise rate and extent of dryland degradation is not known for most parts of the world, but the data presented by the UN are no worse than those offered by its critics. It should be remembered that the UN does not fabricate its own data. It collects them from national experts and governments. Enough is known to conclude that the problem is great, and that it will get considerably worse if nothing is done.

The important thing is for everyone concerned to use their resources and time in finding solutions to the extremely complex and intractable problems in providing sustainable livelihoods for people who live in the drylands. To do that, we must first 'listen to the people'. The following chapter by Solon Barraclough reviews many of the issues that need to be dealt with.

NOTES

The views expressed here are my own and do not necessarily reflect those of the United Nations. I would like to thank Franklin Cardy and Charles Hutchinson for helpful comments and advice, though any errors are my own.

REFERENCES

Aubreville, A. (1949) *Climats, Forêts, et Désertification de l'Afrique Tropicale*. Paris, Société d'Editions Géographiques, Maritimes et Coloniales.

Behnke, R., I. Scoones and C. Kerven (eds) (1993) *Range Ecology at Disequilibrium*. London, Overseas Development Institute.

Dalstead, K. (1988) The use of a Landsat-based soil and vegetation survey and geographic information system to evaluate sites for monitoring desertification. *Desertification Control Bulletin*, 16: 20–26.

Dregne, H. (1983) *Desertification of Arid Lands*. Chur, Harwood Academic Publishers.

FAO (Food and Agriculture Organization) (1993) *Agriculture: Towards 2010*. Rome, FAO.

FAO/UNEP (1983) *Provisional Methodology for Assessment and Mapping of Desertification*. Rome, FAO.

Glantz, M. and N. Orlovsky (1987) Desertification: A review of the concept. *Desertification Control Bulletin*, **14**: 23–30.

Greenland, D.J. and I. Szabolcs (eds) (1994) *Soil Resilience and Sustainable Land Use*. Wallingford, CAB International.

Hare, F.K. (1977) Climate and desertification. In: United Nations (ed), *Desertification: Its Causes and Consequences*. Oxford, Pergamon Press, pp. 63–167.

Helldén, U. (1980) *A Test of Landsat-2 Imagery and Digital Data for Thematic Mapping, Illustrated by an Environmental Study in Northern Kenya*. Rapporter och Notiser, No. 47, Lund, Lunds Universitets Naturgeografiska Institution.

Helldén, U. (1981) *Satellite Data for Regional Studies of Desertification and Its Control*. Rapporter och Notiser, No. 50, Lund, Lunds Universitets Naturgeografiska Institution.

Helldén, U. (1984) *Drought Impact Monitoring—A Remote Sensing Study of Desertification in Kordofan, Sudan*. Rapporter och Notiser, No. 61, Lund, Lunds Universitets Naturgeografiska Institution.

Helldén, U. (1988) Desertification monitoring: Is the desert encroaching? *Desertification Control Bulletin*, **17**: 8–12.

Helldén, U. (1991) Desertification—Time for an assessment? *Ambio* **20**(8): 372–383.

Helldén, U. (1994) Vegetation and desertification: An overview. Paper abstract for the BioResources '94 conference, Bangalore, India, 3–7 October.

Lamprey, H. (1975) *Report on the desert encroachment reconnaissance in northern Sudan, 21 October to 10 November 1975*. UNESCO/UNEP mimeo.

Mabbutt, J.A. (1984) A new global assessment of the status and trend of desertification. *Environmental Conservation* **11**(2): 103–113.
Nelson, R. (1988) Dryland management: The 'desertification' problem. *Environment Department Working Paper* No. 8, World Bank.
Odingo, R.S. (1990a) The definition of desertification: Its programmatic consequences for UNEP and the international community. *Desertification Control Bulletin* **18**: 31–50.
Odingo, R.S. (ed.) (1990b) *Desertification Revisited*. Nairobi, UNEP.
Olsson, L. (1983) Desertification or climate? Investigation regarding the relationship between land degradation and climate in the Central Sudan. *Lund Studies in Geography Ser. A*, No. 60.
Olsson, L. (1985) *An Integrated Study of Desertification*. Lund, University of Lund.
Olsson, L. (1993a) Desertification in Africa—A critique and an alternative approach. *GeoJournal* **31**(1): 23–31.
Olsson, L. (1993b) On the causes of famine—drought, desertification and market failure in the Sudan. *Ambio* **6**: 395–403.
Pearce, F. (1994a) Encroaching deserts are a 'myth'. *The European*, 2 July.
Pearce, F. (1994b) Deserting dogma. *Geographical* **66**(1): 25–28.
Pearce, F. (1994c) Treaty without a cause? *New Scientist* 25 June, p. 5.
Ringrose, S. and W. Matheson (1986) Desertification in Botswana: Progress towards a viable monitoring system. *Desertification Control Bulletin* **13**: 6–11.
Seghal, J., R. Pofali, R. Saxena and C. Harindranath (1991) *India—Land Degradation Status (human-induced) Map*. Nagpur, NBSS and LUP (ICAR).
Stiles, D. (1981) Relevance of the past in projections about pastoral peoples. In: Galaty, J., D. Aronson, P. Salzman and A. Chouinard (eds), *The Future of Pastoral Peoples*. Ottawa: IDRC, pp. 370–78.
Stiles, D. (1984) Desertification: A question of linkage. *Desertification Control Bulletin* **11**: 1–6.
Stiles, D. (1988) Desertification in prehistory: the Sahara. *Sahara* **1**: 85–92.
Stiles, D. (1994) Background paper on Desertification and Vegetation, Bioresources '94 Conference, Bangalore, India, 3–7 October.
Stiles, D. (1995) Desertification is not a myth. *Desertification Control Bulletin* **26**, in press.
Stiles, D. and A. Kassam (1991) An ethno-botanical study of Gabbra plant use in Marsabit District, Kenya. *Journal of the East African Natural History Society and National Museum* **81** (198): 14–37.
Thomas, D. (1993) Sandstorm in a teacup? Understanding desertification. *The Geographical Journal* **159**(3): 318–31.
Thomas, D.S.G. and N. Middleton (1994) *Desertification: Exploding the Myth*. Chichester, John Wiley.
United Nations (1977) *Desertification: Its Causes and Consequences*. Oxford, Pergamon Press.
United Nations (1978) *United Nations Conference on Desertification. Round-up, Plan of Action and Resolutions*. New York, United Nations.
UNEP (United Nations Environment Programme) (1984) *General Assessment of Progress in the Implementation of the Plan of Action to Combat Desertification 1978–1984*. UNEP/GC.12/9, Nairobi.
UNEP (1991) *Status of Desertification and Implementation of the United Nations Plan of Action to Combat Desertification*. UNEP/GCSS.III/3, Nairobi.
UNEP (1992) *World Atlas of Desertification*. London, UNEP and Edward Arnold.
UNSO (United Nations Sudano-Sahelian Office) (1992) *Assessment of Desertification and Drought in the Sudano-Sahelian Region 1985–1991*. New York, UNSO.

Warren, A. and C. Agnew (1988) *An Assessment of Desertification and Land Degradation in Arid and Semi-arid Areas*. London, International Institute for Environment and Development.

2 Social Dimensions of Desertification: A Review of Key Issues

SOLON BARRACLOUGH
United Nations Research Institute for Social Development, Geneva, Switzerland

INTRODUCTION

In early 1993, the United Nations Environment Programme's (UNEP) Desertification Control Programme Activity Centre (DC/PAC) requested the United Nations Research Institute for Social Development (UNRISD) to co-operate in focusing more attention on the social aspects of desertification control. UNEP asked UNRISD to prepare an annotated bibliography based on the literature dealing with social dimensions of desertification. This would be followed by a paper summarizing the principal issues and conclusions emerging from the literature review.

In its initial request, UNEP raised three sets of questions that it hoped would be useful in helping to focus the literature search:

(1) The key land manager in drylands is often a woman (or man) faced with growing food for her or his family. What does she think she needs, to do it in a more sustainable fashion so as to avoid degrading the land? What assistance (support) would enable her or him to do better?
(2) What are the social impacts of drought (specifics please) and desertification/land degradation? (Data are required, not just generalities.)
(3) What facts have been collected on 'environmental refugees'? When does a pastoralist become a migrant become a refugee etc? What proportion of the costs of remedial measures (food aid, humanitarian military intervention, etc.) could be attributed to desertification or unsustainable environment management or anything similar?

Subsequently, in a later communication commenting on the Preliminary Annotated Bibliography, DC/PAC suggested a few additional questions that it hoped the bibliography and review paper would be able to mention. These included controversies about the extent to which desertification may arise from

Social Aspects of Sustainable Dryland Management. Edited by Daniel Stiles

purely natural and not socio-economic factors; the growing body of literature about indigenous knowledge systems and their application to efforts at more sustainable development; and the institutions required to facilitate information flows from the grassroots to governments and donor agencies.

UNRISD sought the co-operation of the International Institute for Environment and Development (IIED) Drylands Programme in this project. Much of the relevant literature was more readily accessible in the United Kingdom than in Geneva and IIED's Drylands Programme already had an extensive bibliography. Yvette Evers of IIED prepared the Preliminary Annotated Bibliography in consultation with UNRISD.

UNRISD, for its part, viewed its responsibilities as being primarily to build on its experience in analysing social development issues from an interdisciplinary broad political economy approach in order to bring them to bear on UNEP's concerns. The data, analyses and contacts made in its ongoing Programme on the Environment, Sustainable Development and Social Change would be particularly helpful.

Background

The 1992 United Nations Conference on Environment and Development (UNCED) included desertification among the major global environmental issues addressed in its programme of action (Agenda 21). Although drought and land degradation had been a concern of populations in dry regions since the dawn of history, desertification became an international issue only in the 1970s. This was largely due to a prolonged drought in the Sahel region of Africa in the 1960s and early 1970s followed by devastating famines in which hundreds of thousands of people and millions of livestock perished. The Sahel famine coincided with the so-called world food crisis of 1972–3 when grain prices in world markets rose steeply while international grain reserves available to meet market demands were largely depleted.

Both the Sahel famine and 'the world food crisis' were widely attributed at the time to droughts aggravated by land degradation and population growth. As will be seen later, these perceptions were, at best, partial and in many ways they were misleading. They did, however, serve to put problems of land degradation in dryland regions (desertification) high on the agenda of several international agencies.

The United Nations World Food Conference was hosted by FAO in Rome in 1974. It resulted in the creation of the World Food Council (WFC) and the International Fund for Agricultural Development (IFAD). The latter was charged with helping the rural poor in developing countries to improve their productivity and incomes, principally by extending loans for rural development projects with these objectives. As a considerable portion of the rural poor were to be found in marginal dryland regions, desertification became one of IFAD's concerns. UNESCO had launched its Man and the Biosphere (MAB)

Programme in 1971; this focused on social and environmental interactions, with special emphasis on dry regions. UNEP, which had been created only following the Stockholm Conference on Human Environment in 1972, prepared the United Nations Conference on Desertification (UNCOD) held in Nairobi, Kenya, in 1977. This conference addressed ecological, technological, social and policy-related dimensions of desertification. It issued a Plan of Action to Combat Desertification (PACD) that was later approved by the United Nations General Assembly. Various other international agencies, intergovernmental organizations, as well as many national governments and NGOs strengthened or commenced programmes designed to deal with land degradation and its social impacts.

These multiple initiatives in answer to international concern about desertification had not generated any firm consensus in either the scientific community or in political circles about its scale, causes or alleged social and environmental impacts. There was also little consensus on what should be done about it. In fact, many of the widely held assumptions of the 1977 UNCOD conference were being increasingly questioned (Rhodes, 1991). Most of these assumptions were reiterated by UNCED in 1992.

Objectives and content

The purpose of this chapter is to review some of the principal social issues underlying differing perceptions of desertification, its causes, impacts and control. It is based on a very partial review of the abundant literature generated during the last few decades. The paper is divided into three parts.

I. DESERTIFICATION ISSUES: THE VIEW FROM ABOVE

Since desertification problems were placed on the international organizations' agendas in the early 1970s, several issues related to its social dimensions have been the focus of considerable debate and research. These include the definition of desertification, its extent and the numbers of people affected, its causes and its consequences.

The concept of desertification

One reason for the lack of consensus about desertification is the ambiguity of the concept. The term is much more useful for mobilizing political support to combat what are imagined to be desert sands marching over once-fertile crop lands and productive pastures than for analysing the causes, effects and proposed remedial actions to deal with the multiple processes generating land degradation. These processes may often be interrelated and mutually reinforcing but they are always to some extent specific for particular sites and times.

UNEP has defined desertification as 'a complex process of land degradation in arid, semi-arid and sub-humid areas resulting mainly from adverse human impact' (UNEP, 1992). UNCED broadened this definition to 'land degradation in arid, semi-arid and sub-humid areas resulting from various factors, including climatic variations and human activities' (UNCED, 1992). Some authorities claim to have found over a hundred definitions of desertification (Rhodes, 1991; Barrow, 1991).

Differing definitions of what is supposed to be an operational concept stem from divergent values, experiences, interests and objectives. Each definition of a process such as soil erosion carries with it differing ideological nuances (Blaikie, 1985). Different definitions of desertification also frequently imply divergent political agendas. For example, UNCED's broadening of UNEP's definition of desertification to give more emphasis to climatic change suggests the possibility of developing country governments seeking an additional rationale for requesting rich countries to bear more of the costs of its negative impacts and of its control. Of course, this would also depend upon the extent to which climatic change could be linked to greenhouse gas emissions that for the most part originate in rich industrial countries (Toulmin, 1993).

Most definitions of desertification agree that it implies land degradation in dry regions. Dry regions can, at least theoretically, be rather precisely delimited by relating potential evapotranspiration to amounts and patterns of rainfall (UNEP, 1992; Barrow, 1991). The practical difficulties of doing this accurately for a given place and time period are nonetheless extremely formidable.

Land degradation is an elusive concept even at theoretical levels. It implies a lessened capacity of the land to produce. Net degradation is the difference between degradation from both natural processes and human interventions, on the one hand, and restorative natural and human processes, on the other (Blaikie and Brookfield, 1987).

Production and productivity, however, are socially defined.[1] Land degradation is a social concept, as is desertification. Hunter–gatherer societies will have different perceptions of land degradation than those of peasant agriculturalists. Members of both peasant and hunter–gatherer societies will perceive degradation processes differently from commercial farmers and other land managers in industrial societies. Within the same society, perceptions may vary greatly according to the observers' experience, class position, social status, gender and many other factors. Definitions and measurements of land degradation are necessarily to a large extent arbitrary.

The issue becomes even more complex when different scales of geographic areas and of time are considered. Eroded soil from a farmer's field may be deposited by wind or water on other fields that may benefit someone else. Net degradation from erosion tends to decrease as the size of the area being analysed increases (Blaikie, 1985). On the other hand, losses in productive capacity due to reduced biodiversity or degraded vegetation will likely be less serious when restricted to limited areas than when they affect extensive regions. Moreover, some lands that are degraded by drought and by inappropriate human activities may bounce back to their previous productive potential rather quickly when these factors are eliminated. In other cases, however, recovery may require prohibitively costly investments or recuperation periods of decades and possibly millennia. These considerations lead several analysts to reject the term of desertification as an analytical category and to introduce resilience—the ability of the ecosystem to return to its former state following disturbance—into their concepts of land degradation (Nelson, 1990; Horowitz, 1990; Hammer, 1993; Toulmin, 1993).

In summary, it is impossible to find a definition of desertification that permits a consensus to emerge on how it should be measured, compared and monitored across differing ecological and social systems and through various periods of time. The identification and measurement of land degradation is partly an ideological and political issue. Nonetheless, some concept of desertification has to be adopted and made operational in order to discuss its social dimensions. This discussion attempts to follow the UNCED definition cited above.

The extent of desertification

Accepting UNCED's definition of desertification, how extensive and serious is land degradation in dryland areas and what are the trends? The review of the literature suggests considerable differences in opinion about answers to these questions. These disagreements are in part a corollary of the conceptual issues discussed above. They are also in part a result of practical measurement and monitoring deficiencies. In some cases, differing estimates may even stem from careless arithmetic.

Areas affected

Estimates by Dregne and by Mabbutt in the early 1980s suggested that over 30 million km^2 suffered from at least moderate desertification (Dregne, 1976; Grainger, 1990). This amounted to about one fourth of the earth's land area and over two thirds its dryland areas, excluding hyper-arid deserts. Most of these degraded drylands were in Africa and Asia and were rangelands. The proportion of drylands at risk that were already suffering at least moderate desertification were believed to range from about 70% or 80% in Africa, Asia and South America, to less than half in North America and Australia. Desertification was estimated to be increasing at about 200 000 km^2 annually (Grainger, 1990; Barrow, 1991).

There are many problems with these estimates when it comes to assessing the social dimensions of desertification. Only about 30% of the land area believed to be degraded was attributable to soil degradation (UNEP, 1991, 1992; Toulmin, 1993). The remaining 70% was attributed to lands where degradation in vegetation was not accompanied by soil degradation.

Degradation of vegetative cover implies replacement of 'climax' vegetation by other less desirable plant associations (Grainger, 1990). Debates about the meaning and significance of climax vegetation have continued among ecologists ever since the concept was introduced. Did bison and their Indian hunters degrade great plains' vegetation in North America? If so, what was the climax cover? How can one determine when 'natural' vegetation has been replaced by 'inferior' plant associations with lower productive potential, for example? Was it always land degradation to replace woodlands in dryland areas with crops and pastures? In many circumstances clearing forests to make way for crops or other uses may represent the best economic choice for society. Moreover, agricultural use may be potentially sustainable in many formerly wooded dryland areas given appropriate management practices. Are native shrubs always more productive than annual grasses that may be more useful for pasture? The conceptual and practical difficulties of estimating degradation of vegetative cover seem to be even more formidable than estimating degrees of soil degradation (Toulmin, 1993; Horowitz, 1990).

Inclusion of areas where vegetation but not soils are believed to have been degraded increases the areas of drylands estimated to be affected by desertification from about one fifth of the total area to over two thirds of the total. It also raises many additional issues in analysing the social dimensions of desertification. These global estimates are extremely vulnerable to criticism as they confound processes that are clearly land degrading with others about which there is a great deal of uncertainty concerning their long-term impacts on land productivity. Including degraded vegetative cover on non-degraded soils in estimates of areas affected by desertification may be counterproductive for mobilizing donor support for programmes to improve natural resource management in dryland areas (Toulmin, 1993).

Many of these ambiguities are evident in UNEP's *World Atlas of Desertification* (UNEP, 1992). The atlas was prepared under the direction of a group of world eminent experts. Its maps and tables showing the extent of drylands by region and the proportions that have suffered various degrees of soil degradation appear to be the most authoritative to date. But the preface discussing the extent of desertification in various regions cites estimates of total dryland areas that have been degraded, including those where only vegetation was affected. The preface states: 'In terms of severity of [dryland] degradation, however, North America and Africa are by far the worst off with 76% and 73% of their drylands degraded'. This could only be plausible if most changes in dryland vegetation in North America during the last few centuries were considered to be degrading. This is a debatable question as only about one tenth of drylands soils in the region were estimated in the *Atlas* (table 2) to have been degraded as compared with nearly twice as high a proportion in Africa. UNCED's Agenda 21 estimates that worldwide some 3.6 billion hectares of drylands have been degraded. The atlas tables suggest, however, that soils have been degraded (at least slightly) on only about 1.04 billion hectares of drylands.

Number of people affected

The problems of estimating the populations affected by desertification are even more intractable than those of assessing extents and rates of land degradation. In a global economy, practically everyone is affected directly or indirectly by any significant change in global agricultural markets whether induced by desertification or something else. If desertification is affecting agricultural output, costs, migration rates and the capacity to resist droughts significantly on a worldwide scale, then all the world's people may eventually be affected in one way or another.

Estimates made by Mabbutt and also by Dregne in the early 1980s suggested that nearly 300 million rural people and another 200 million urban dwellers were directly affected by desertification. These earlier data were evidently based on estimates (often highly notional) of the number of people residing in dryland regions suffering from land degradation (whose areas were extremely problematic). UNCED's 1992 figure of 900 million people in dryland areas at risk from desertification was derived by combining FAO's estimates of the agricultural populations in sub-national administrative districts with UNEP's maps of dryland regions in danger of degradation, according to information provided by a UNEP official.

Even assuming away problems of estimating the numbers of people residing in areas at risk from desertification, the question remains of how they are affected. Most may experience negative impacts on their livelihoods, as the literature usually assumes, but the effects are undoubtedly different for divergent groups. In the past, some individuals and social groups probably benefited, in several respects at least, as there are usually winners and losers in any process of change in socially differentiated societies. Until now at least, some social groups have occasionally benefited from desertification at the expense of others in terms of increased political power, wealth and income. For example, large landlords and speculators were able to appropriate the lands of drought-stricken peasants in Brazil's north-east for virtually nothing, while the former had to offer their labour to local or other employers at starvation wages (CIDA, 1966b). Much of the literature reviewed virtually ignored this contradictory reality at global and regional levels, although there were exceptions (Blaikie, 1985; Blaikie and Brookfield, 1987). Several local level case studies, however, attempt to consider who gains and who loses, as will be seen later.

Conclusions

The extent of land degradation in dryland regions is very great, and is spreading. Lumping together soil degradation of various kinds with areas suffering only vegetative degradation under the blanket term of 'desertification' is confusing. It is unhelpful for promoting a better understanding of the severity of the problem or of its causes, impacts and possible remedies. This is especially the case as the alleged negative effects of many changes in vegetative cover considered to be degrading by some analysts are extremely controversial.

Similarly, the estimate of 900 million people at risk from desertification is not particularly useful for appreciating the social implications of dryland degradation. These represent only the agricultural (including pastoral) populations residing in dryland areas where soils and vegetation are believed to be in danger of becoming increasingly unproductive. It does not include urban and other non-agricultural residents in these areas who also may be very directly affected by land and water degradation, nor does it include populations outside these dryland regions who may be seriously affected directly or indirectly by externalities such as water scarcity, sediments, floods, decreasing biodiversity, higher prices, migrations, social conflicts and the like. The estimate does not take into account the fact that different social groups are likely to be affected in divergent ways, both negatively and, by at least some criteria, positively.

Causes of desertification

For analysts who use a concept of desertification attributing it primarily to human activities, its causes are mainly social by definition. There is usually discussion of an assortment of direct and indirect social causes, but natural processes such as long-term geological, biological and climatic evolutions or fluctuations that may be largely independent of human activities are usually relegated to the background context. On the other hand, analyses of desertification that start from a definition emphasizing both natural causes, such as most major droughts in the past, and human activities in degrading drylands deal more explicitly with interactions between natural and social systems and subsystems.

Interacting natural and anthropogenic causes

UNEP's attempt to ascribe the main direct causes of all soil degradation in each region to different human activities such as deforestation, overgrazing and non-sustainable agriculture (UNEP, 1992) is easily understandable because these processes may be directly influenced by public policies. It is misleading, however, as it ignores natural causes of soil degradation interacting with human interventions. Examples would include most past climate changes, geologic uplift accompanied by accelerated soil slippage and erosion, many insect and other plagues, volcanic eruptions covering large areas with ash while also influencing climate, some naturally caused fires and the like.

Oversimplistic approaches to the causes of land degradation have been widely criticized by several specialists (Glantz, 1987, 1988, 1989; Spooner and Mann, 1982; Blaikie, 1985; Blaikie and Brookfield, 1987). The relationships between drought, land degradation and society, for example, differ in each situation. Enough is known, however, to be able to predict that areas subject to severe droughts and longer-term climatic cycles in the past will probably continue to experience similar fluctuations in the future, irrespective of human activities.

Droughts often stimulate sequences of actions and reactions leading to long-term land degradation. They may also trigger local food shortages, speculation, hoarding, forced liquidation of livestock at depressed prices, social conflicts and many other disasters associated with famines. These may catastrophically affect numerous groups and strata of local populations (Frankie and Chasin, 1980).

Pastoral and peasant societies learned how to cope with the risks of drought after centuries of trial and error. These strategies allowed the group to survive although usually with considerable hardship for many of its members. Traditional coping strategies were disrupted by incorporation into expanding world markets accompanied by growing monetarization, commercialization, colonial and post-colonial land alienation and taxes. These disruptions continued with the conflicts, policies and institutional changes associated with the creation of independent national states bent on 'modernization'. Droughts alone can seldom be blamed for accelerating desertification or for the hardships accompanying famines. Both desertification and famines are the outcomes of interacting social and natural systems locally, nationally and globally (Glantz, 1987).

A good case in point was the 1968–74 Sahel drought mentioned earlier. Research during the 1970s showed that social processes, institutions and public policies, within the context of pre-drought international economic and political relationships, provided much better explanations of the famine and its impacts on different social groups than did drought alone. These other factors were also largely responsible for the long-term land degradation that took place. Drought was merely an important catalyst (Frankie and Chasin, 1980).

The International Federation of Institutes for Advanced Studies (IFIAS) sponsored exhaustive research on causes of the Sahel famine as part of its study of the role of climate changes in producing the 1972–3 'world food crisis'. IFIAS's project 'Drought and Man—The 1972 Case Study' concluded, after a great deal of exhaustive research, that drought in various parts of the world in the late 1960s and early 1970s was only a minor factor contributing to the 'world food crisis'. Socio-economic and political factors such as an unstable world monetary system and changes in some agricultural policies in the United States and USSR had been far more important. The IFIAS study of the Sahel famine reached similar conclusions. Neither the famine's root causes nor its social impacts could be primarily attributed to drought or to desertification although these were contributing factors (Garcia, 1984; Garcia and Escudero, 1982; Garcia and Spitz, 1986; Independent Commission on International and Humanitarian Issues, 1985; Barraclough, 1991a).

The Sahelian case illustrates the complex local, national and international socio-economic and political factors underlying the frequently catastrophic

Photograph 2.1. The 'world food crisis' of the early 1970s, which led to massive food relief efforts in Africa, was due more to socio-economic and political factors than to drought and desertification. (Photo: UNEP/ICCE)

impacts in underdeveloped countries of drought and land degradation. In seeking to understand the causes of desertification, however, it is instructive to look at the historical record for particular sites and regions. This is no easy task, as written historical records seldom address most of the relevant questions, and in any event they do not even exist for many places and seldom cover long enough periods to assess secular changes in climate. Digging up and analysing sufficient complementary archaeological, geological, paleoecological and paleoclimatical data is costly and arduous. Where it has been attempted, there are always pieces of evidence missing and many contradictory hypotheses. Nonetheless, some tenuous conclusions about the role of natural and social processes in land degradation can be drawn for a few places where such studies have been made. These support what has been said above about interacting natural and human processes.

One of the earliest agricultural-based civilizations, the Akkadian empire in Syrio-Mesopotamia that had prospered between 2600 and 2200 BC, apparently collapsed as a result of drought-induced desertification. This society had reached hitherto unmatched development and prosperity by 2200 BC. It embraced an area spanning more than 1200 km from the Persian Gulf to the headwaters of the Euphrates river with over 100 000 inhabitants and numerous cities when it suddenly disappeared. The empire's fall was apparently triggered by an abrupt climate shift that left the region exceedingly dry for over 300 years. Traditional agriculture could no longer support the cities and drought-stricken migrants invading from further north soon overwhelmed them. The empire's highly centralized administrative system of control and surplus extraction undoubtedly contributed to its vulnerability to the onset of drought. The dry spell was associated with increased volcanic activity in Anatolia, but massive warming of ocean currents disrupting local weather patterns seems to be a more likely explanation of the long dry period. In any case, the climatic change was unlikely to have been induced by human activities (Weiss *et al.*, 1993).

Blaikie and Brookfield examined evidence from the sub-humid Mediterranean region and from humid north-western Europe. They wrote: 'We do not reach any firm conclusions, indeed cannot, but are able to suggest that variability in the "human" elements in the equation can be given some degree of precedence even though natural events outside the present normal range of experience may also have been significant.' They emphasize three general conclusions. First is the importance of land management in relation to crops and the land-use systems. Second are the linkages between the ability of farmers to manage land sustainably and the conditions of the state and its economy. The latter circumscribe the possibilities of the former. For example, social transformations at the state level may facilitate more productive and sustainable land management even with increasing population pressures on resources, while in other cases such increased population pressures could spell disaster. Third, they stress the importance of occasional extreme events:

> The point that emerges is not that a single disaster creates degradation, but that a succession of disasters has a particularly damaging effect when ongoing social and economic conditions are such as to expose the production system and the land to abnormal harm from such events . . . This conclusion also points up the potential linkage between the study of land degradation and that of natural hazards and their impact . . . (Blaikie and Brookfield, 1987).

Numerous other studies suggest similar conclusions. For example, overgrazing can lead to land degradation in some situations, but so can undergrazing in others (Nelson, 1990; Horowitz, 1990). High population densities and intensive cropping frequently lead to land degradation. In the pre-conquest Andean region, however, growing population pressures were accompanied by careful terracing and water management with sustainable high levels of land productivity. Following the Spanish conquest, the population decreased drastically. Terraces and irrigation systems were neglected while land degradation greatly increased (Treacy, 1989). There are also numerous recent examples of where decreased rural populations have been followed by severe land degradation when sustainable intensive farming systems decayed (Garcia Barrios and Garcia Barrios, 1992; Barraclough and Ghimire, forthcoming). Each case is in many respects unique and has to be analysed within its own particular context.

But such analyses should not be limited to local levels. Broader ecological, socio-economic and political contexts also have to be explicitly taken into account and integrated into the analysis.

Direct and indirect causes

Much of the desertification literature reviewed makes at least an attempt to follow this dictum about taking into account broader contexts. For example, Grainger (1990) concludes his chapter on causes of desertification as follows:

> Desertification has four direct causes, overcultivation, overgrazing, deforestation and mismanagement of irrigated cropland. These do not occur by accident, but are greatly influenced by the effects of growing populations, economic development and conscious policy decisions by governments and aid agencies.

Grainger also mentions other factors such as cash crop expansion, land tenure, poverty, underdevelopment and trade policies. The problem is that he makes no analysis of social systems and of how they are functioning in specific contexts. There is a listing of 'direct and indirect causes' of desertification, but no integrated analysis of relationships and hierarchies among them. The probability that similar social processes and policies might lead to very different outcomes in different social contexts is hardly considered. There is little recognition that systemic changes are highly unpredictable in both their occurrence and their effects, but that they happen rather often in both natural and social subsystems.

This simplification of the issues permits Grainger to dedicate the last half of his book to discussing measures designed to control desertification. These, for the most part, are intended to attack its direct causes. He compares desertification to a chronic disease that demands continued medication (Grainger, 1990). Anyone reading Blaikie and Brookfield, Spooner and Mann or Glantz might justifiably question whether the medication Grainger proposes may not be similar to treating malaria or AIDS with aspirin. In fact, he recognizes this in his concluding chapter.

Conclusion

Natural and social processes both play important roles in causing desertification. They are usually intextricably intertwined, but in different mixes in each local situation. Among natural causes, drought is the most important in stimulating sequences of actions and reactions that frequently are accompanied by land degradation. In some circumstances, however, droughts may contribute to the emergence of social strategies that enhance sustainable land productivity while protecting local livelihoods (Blaikie and Brookfield, 1987).

The role of human activities in influencing the frequency, extent and severity of droughts is too complex to be treated in any depth here. It is sufficient to note that there is clear evidence that in some cases they are crucially important in accentuating local climate variations while in others they seem to have had only a minor role. The evidence that human activities influence global climate patterns through greenhouse gas emissions and in other ways is also compelling. The uncertainties inherent in global climate modelling, however, do not permit firm predictions. There is a great deal of uncertainty concerning human influence on climates in the recent past and their possible effects in the future (Schneider, 1989; Glantz, 1984, 1987).

The social processes and structures contributing to desertification, or to its control, are infinitely varied at local levels. Facile generalizations are always plagued by multiple exceptions. The so-called 'direct causes' such as overgrazing, careless cultivation, overuse of irrigation, deforestation and the like cannot by themselves take one very far in understanding the social dynamics leading to land degradation, as these are usually generated by much broader social processes.

Explanations of desertification assuming peasant ignorance and short-sightedness were especially in vogue with colonial administrators. These have been largely discredited by research illuminating traditional peasant farming and social systems and the processes disrupting them such as land alienation, surplus extraction and commercialization. On the other hand, explanations assuming short-term profit-maximization goals by commercial farmers who can escape their costs of land abuse by shifting them to society are also incomplete. Non-sustainable land-use practices by well-off commercial operators as well as by poor peasants and herdsmen are, for the most part, determined by policies and institutions over which they have little or no control. Among these, national and international economic and political relations are of paramount importance.

More effective control of desertification requires a better understanding of the processes and socio-economic relations generating it at all levels. It also requires perceptions by those wielding economic and political power locally, nationally and internationally that their own interests are at stake.

Social impacts of desertification

Mention has already been made of possible global impacts of land degradation through its influence in stimulating social tensions, migrations and market shortages of some important commodities produced in dryland regions. Also, it may contribute to climatic change, although this remains controversial. One must treat quantitative estimates of these impacts with great caution as all are based on questionable assumptions.

Costs of desertification

Estimates of several tens of billions of dollars' annual loss from land degradation, for example (Barrow, 1991), seem rather meaningless.

Comparing monetary costs and benefits across different social systems with divergent values presents insuperable epistemological difficulties. Land degradation implies lower land productivity by definition. Its economic and social significance is likely to be very different for largely self-provisioning peasants in Nepal or Tanzania than for commercial farmers in Western Europe, Japan or the United States. In OECD countries, governments are spending a total of well over $100 billion annually in agricultural subsidies largely to protect their agricultural producers from the ravages of falling prices accompanying market gluts (UNCTAD, 1991).[2]

One of the biggest difficulties in assessing social impacts of desertification is to separate those attributable to land degradation from those arising from other processes associated with desertification, or even those rather independent from it. Drought and longer-term climate changes, land alienation and other ravages of 'modernization', wars and other social conflicts, structural adjustment and related public policies, and demographic changes are only a few examples.

Desertification is frequently only one factor in explaining the social disruptions and social costs for which it is sometimes blamed. Famine in the Sahel discussed above and severe poverty with massive out-migration in Brazil's north-east mentioned below are only two examples. On the other hand, the

Photograph 2.2. Some ecologists predict catastrophe as society approaches its limits to economic and demographic growth. These low-caste Indians have little chance of a better life due to social and economic limitations. (Photo: Daniel Stiles)

longer-term costs to society of land degradation may be much greater than perceived even by some observers who are considered to be alarmists. This is primarily because the options for unborn generations may be seriously curtailed. How important this appears largely depends on one's vision of the world and its future. Many economists foresee infinite human capacities for social and technological adaptations and innovations (World Bank, 1992; Simon, 1981; Beckerman, 1974, 1992). Some ecologists and other scientists, however, predict catastrophe as society approaches its limits to economic and demographic growth (Erlich and Erlich, 1970, 1992; Brown, 1973, 1979; Meadows *et al.*, 1970; Wilson, 1988). These are not issues that can be fully resolved by empirical research. The questions they pose remain, in some measure, metaphysical.

Costs for whom?

A more immediately practical issue about the social impacts of desertification is that of costs and benefits for whom? This was mentioned earlier in connection with the discussion of populations affected. Some groups benefit and others lose, although even those who gain in wealth and power may be negatively affected by a poorer environment. Within these groups, some sub-groups and individuals benefit or lose much more than others.

Subsuming the benefits and costs for different groups and individuals within some global cost-benefit calculus involves heroic assumptions that are questionable philosophically and likely in any event to be misleading as a basis for action. Scrupulous analysts of the social impacts of desertification should attempt to identify explicitly how different social groups and classes are affected before imposing their own choices of values for estimating social costs and benefits. This would allow others with different values to make their own calculations. Also, it would facilitate identification of policy interventions aimed at specific problems.

A BOTTOM-UP APPROACH TO THE SOCIAL DYNAMICS OF DESERTIFICATION

To approach desertification issues starting from the perspectives of those subsisting on lands in vulnerable dry regions is no easy matter. Some 900 million people gain their living directly from these lands while many more are supported by them indirectly. For the most part these people depend crucially on dryland productivity for eking out bare livelihoods, although some may be relatively well off. A few may even be large and influential producers with good access to government credits, services, protection, infrastructure, modern technology and sheltered markets. In most poor countries, peasants and pastoralists are competing among themselves as well as with powerful private and state corporate interests for the same scarce land and water resources.

In each locality different combinations of institutions regulate access to resources and the divisions of labour and of production. Local populations are always socially stratified in a variety of occupational and status roles. Lineages, ethnic or religious identification, caste, class, age and gender as well as linkages with markets and political authorities outside the local communities all influence what social groups are most relevant to consider in a given locality when trying to understand livelihood issues and their relations with land management. Moreover, many rural societies in developing countries are undergoing violent disruptions as they become increasingly incorporated into regional, national and international markets, power structures and conflicts. Traditional social relations and livelihood systems are frequently overridden, deformed or crushed with shocking brutality. These traditional systems were often cruel and very inequitable, but all too often what replaced them was even worse for many participants and for the environment.

In opting for a bottom-up approach in attempting to understand better the processes and relationships generating land degradation and determining their social implications, complications such as those just mentioned have to be kept constantly in mind. At the same time, it is necessary to simplify in order to relate what is happening at local levels to broader national, regional and global possibilities and trends. This implies finding in-depth local-level case studies that bring out interactions. Cases should represent several different important patterns of social institutions, ecological contexts and processes of socio-economic and political change taking place in regions vulnerable to desertification. The case studies should illuminate the local-level dynamics that induce different individuals and social groups in their pursuit of livelihoods or profits to behave in the ways they do. Constructing such a typology is a major challenge.

Ideally, local-level case studies should also bring out how each group's alternatives and incentives are constrained by existing social arrangements and what opportunities may exist for relaxing these constraints. These questions cannot be restricted to analyses of local communities. Institutions, policies, market forces and conflicts at national and international levels sharply constrain vulnerable social groups' quests for survival and improved livelihoods. As different social groups have conflicting as well as complementary interests in their access to resources, conflicts and potential conflicts have to be taken into account in analysing alternatives. Case studies of this kind imply that the problems and perceptions of the major social actors have been understood and reported.

This agenda is utopian, but it suggests what to look for in reviewing case study materials in the literature. The rest of this section attempts to follow this bottom-up approach in examining social dimensions of desertification.

Local-level perceptions and dynamics

The case studies reviewed make clear that the negative social effects associated with land degradation are felt principally by the rural poor. Better-off groups and strata, by definition, have better access to resources, community services, markets and infrastructure. They also have more access to political power. The most vulnerable groups usually include the landless and near-landless, pastoralists with lower status or smaller herds, ethnic or religious groups who, while not necessarily minorities, are subordinate, and refugees. Within these less privileged groups, the elderly, women and children are likely to be particularly vulnerable, although this is not always the case. These generalizations verge on being tautologies. Wealth and power tend to be nearly synonymous in most societies as do poverty and powerlessness.

Local social systems are highly diverse as are the farming systems accompanying them. By the late twentieth century, practically all local societies have become incorporated into the world industrial–commercial system, but in varying ways and degrees. Each situation has to be studied with the participation of the various social groups involved to determine who the major actors are, what their principal concerns may be, what roles they are playing and why, as well as if there are any feasible alternatives.

This is an important conclusion. Those proposing national or international programmes to combat desertification or to alleviate its social effects should heed it seriously. The popular NGO admonition of 'Think globally but act locally' should always be supplemented by its corollary of 'Act globally but think locally taking the whole wide range of local conditions into account'.

Who are the land managers?

The concept of local land managers has to be re-examined in concrete situations in order to be operational. It implies an economic unit such as a peasant holding, family farm, or a large ranch or plantation with a centralized management. The family heads or the managers of large units are assumed to make land-use decisions as well as those concerning farming practices, investments, self-provisioning, marketing and the like. This is a long way from reality in many contexts.

In customary farming systems in West Africa, for example, different members of the same extended family group living in the same compound make various production, consumption, investment and land-use decisions responding to traditional inheritance and other social rules. Their agricultural and pastoral activities take place in organizational contexts that have little relation to the Western concept of a farm unit (Savané, 1986, 1992; Barraclough, 1991a). The same is true for sub-units of latifundia-minifundia systems in Latin America, and surviving customary communal systems in both Latin America and Asia (Jodha, 1982; Barraclough, 1991a).

When the male heads of peasant families were persuaded by state development agents sponsored by the United Nations Andean Community Development Programme to plant trees on abandoned idle lands near Lake Titicaca, for example, the trees were ripped up by their womenfolk soon afterwards.

The 'abandoned' lands were part of the women's long-term brush–fallow–potato rotation (CIDA, 1966a). In Brazil's semi-arid north-east, peasants frequently have no secure tenure rights. Their land-use and cropping practices are practically dictated by absentee landlords, merchants or moneylenders (Johnson, 1973; CIDA, 1966b; Barraclough, 1991b).

The historical context

Each local situation affected by desertification has to be understood in its particular historical perspective. For example, the arid and semi-arid Algerian steppe includes 20 million hectares and supports some 6 million head of livestock, mostly sheep, raised by nomadic steppe dwellers. Before European colonization in the nineteenth century, these nomads moved their flocks throughout the steppe in accordance with the seasons and climate changes to take full advantage of its pasture while minimizing risk. They also had complementary trade and land-use relations with settled farmers in the north and oasis dwellers to the south. These inter-group relations included barter of many vital products and seasonal grazing rights for all three groups on some lands controlled by the others to make the greatest use of available scarce water, rangeland and cropland resources. This traditional pattern broke down under French colonial rule. Colonists appropriated the agricultural lands of the north for modern commercial export-oriented agriculture. They excluded the nomads while the colonial state closed access by the nomads to remaining forest. It also imposed taxes.

Social inequalities and land degradation both increased rapidly as did pauperization of a majority of the nomads, although a few became large breeders, and the virtual extinction of their traditional culture. The war for independence led to further social disruption as did many policies of the independent state later. Government efforts to establish 'green belts' and 'pastoral co-operatives', while apparently well intentioned, were accompanied by further social and ecological degradation. Out-migration of nomads to the cities accelerated, where most continued in poverty. Desertification of the steppe and the negative social effects accompanying—and causing—it cannot be understood without reference to this historical background. Efforts to rehabilitate the area and its people will inevitably fail if they are not based on a knowledge of the historical reality and do not enjoy the effective participation of the groups most affected (Bedrani, 1983).

This case is summarized briefly here because its sensitive treatment in a dynamic historical context of major socio-economic, political and ecological processes associated with desertification makes it one of the better ones encountered in the literature review.[3] It shows why the social effects of desertification cannot really be separated from its causes and why 'ecological refugees' can seldom be neatly distinguished from refugees fleeing the ravages of war or 'modernization'. In one sense, a principal cause of desertification and its negative social impact during both the colonial and post-colonial periods was what is commonly called 'development'. This included 'modernization' of commercial agriculture where profitable, urbanization and, especially after independence, industrialization. But the ways these development goals were pursued disrupted ecologically sustainable land-use systems while excluding increasing numbers of nomads and other groups from their traditional sources of livelihood before alternative opportunities for gaining a living were available for them. Clearly, both the content of development and the strategies for achieving it will have to change profoundly in the future if it is to be sustainable either socially or ecologically in regions such as the Algerian steppe.

Social impacts of drought

Droughts affect different social groups in divergent ways. For example, a study carried out among Beja famine migrants in the eastern Sudan in 1985 can be summarized as follows:

- The semi-nomadic Beja households were very dependent upon livestock for their incomes and consequently were unwilling to sell the majority of their animals even though most of them were dying of starvation and diseases.
- The prices of livestock which were sold had fallen to approximately one quarter of the pre-drought levels by January 1985.
- At the same time, the price of the staple grain, sorghum, had risen to six times its pre-drought level.
- Households were forced to rely on self-employment to a far greater degree than had previously been the case; woodcutting, charcoal burning and mat-weaving seem to have been the most widely practised means of self-employment.
- An increase in labour migration from nearby villages also occurred.
- Access to traditional lines of credit through merchants collapsed as collateral livestock disappeared.
- Mortality and morbidity among households who had migrated to the roadside were high, mainly because of the inadequate provision of public relief to the victims.

There were marked differences in responses to drought between predominantly peasant and predominantly pastoralist producers. Pastoralists appeared to be generally unwilling to sell or slaughter livestock, whereas those households which relied mostly on crop production for food were likely to be more willing to part with their animals. Pastoralists were so dependent upon their camels for transportation that these animals were sold only when all other assets had been exhausted or were left with relatives. The propensity to sell camels appeared to be particularly low for the Beja, indicating that petty trading was an important fallback strategy for this tribe.

But this is only one case. A great many others in different contexts indicate somewhat divergent outcomes. Great care has to be taken in making generalizations. The victims are likely to attribute their plight to drought in answer to superficial questions. Deeper probing, however, usually reveals a host of other factors.

Ecological refugees

Large flows of refugees in Ethiopia, Somalia and the Sudan among many other places have been associated with droughts, falling land productivity and armed conflicts (Hjort and Salih, 1989; Hutchinson, 1991; Johnson and Anderson, 1988; Markakis, 1993; Keen, 1992; Allen, 1993; Buchanan-Smith and Petty, 1992; Davies *et al.*, 1991). They were also associated with the expansion of large-scale commercial agriculture as well as with massive land alienation for export production, 'development' projects, parks and game reserves, and the like. Moreover, they were in part due to numerous government policies that discriminated against weaker social groups and favoured stronger clients of the state. Additional factors were changing international terms of trade, foreign 'aid', together with political and military support for governments or their opponents. They also frequently accompanied changing demographic patterns and increasing numbers of rural people. It is quite impossible in most cases to say what weight should be given to each of these many factors causing peasants or pastoralists to become refugees.

Refugees from the 1968–74 Sahel drought were surveyed in Burkina Faso, Mali, Mauritania and Niger in 1973 by the Center for Disease Control. One of the key findings of the survey was that to cope with the loss of water and pasturage nomads attempted to carry out their traditional drought-coping responses in a different way from that which had been customary. Entire families were migrating south to areas not usually part of the nomads' territory and to urban areas. Furthermore, the nomads were migrating for reasons other than to search for pasturage. Some groups were migrating solely to reach food distribution centres.

These unusual migrations caused many social and political problems both at the national and sub-regional level. Major Sahelian towns found that their populations suddenly grew by 50% (Mopti, Mali), 42% (Dakar, Senegal), and 66% (Noakchott, Mauritania). Rosso in Mauritania had the largest increase in population as a result of the drought, 94%. Border towns of neighbouring countries were also confronted with a sudden influx of refugees. Fifteen per cent of the rural population of Senegal migrated to other areas. In northern Mauritania, because of high cattle mortality rates, 80% of the inhabitants moved to the larger towns in the south.

This rapid influx into the larger towns and capital cities posed political and economic difficulties for the African governments which were ill-equipped to handle the situation. Relations between states were strained as affected popu-

Photograph 2.3. Ecological refugees and migrants are created by a number of inter-related environmental, socio-economic and political causes. (Photo: UNHCR)

lations moved to other countries. Migrants from Mali crossed into Benin, Burkina Faso, Niger and Nigeria. The CDC survey reported 40 000 Malian Tuaregs crossing the border into Niger. Urban unrest in 1971 and 1973 in Senegal was partly attributable to the drought and its effects. Governments sometimes manipulated the disaster to settle old political scores with opposition groups. Those of Mali and Chad are said to have used the drought to break the strength of nomadic groups by withholding aid (Somerville, 1986).

These migrants can be considered environmental refugees in the sense that drought triggered their departure. There were a great many other social processes and institutions at play, however, that would probably have caused many of them to migrate sooner or later in any event (Frankie and Chasin, 1980). The same was seen to be the case for migrants fleeing droughts in north-east Brazil discussed earlier. The concept of 'environmental refugee' is extremely difficult to pin down operationally without making a large number of rather arbitrary assumptions.

In particular cases, a single factor such as drought or civil conflict appeared to be the principal catalyst in turning peasants, pastoralists or workers into refugees. More in-depth analysis usually shows that many other factors also were important contributing causes to their flight (Gray and Kevane, 1993; Fuller, 1987; Ibrahim, 1982; Buchanan-Smith and Kelly, 1991; Allen, 1993; Keen, 1992).

This leads one to question the utility of the concept of 'environmental refugees' for analytical purposes. The distinctions among causes of people being refugees is often politically useful for mobilizing government action and international assistance. These distinctions among refugees are also required for administering certain kinds of relief programmes and development projects. Aid administrators and immigration authorities may have to distinguish 'political' from 'economic' or 'environmental' refugees, for example. These distinctions, however, are necessarily rather arbitrary except for extreme cases. They are primarily administrative conveniences and not social science analytical categories. Social scientists, however, can help administrators in specific cases to design criteria that are as realistic and fair as possible.

Local-level perceptions

The principal issues associated with desertification are likely to appear rather different to the vulnerable groups most affected in rural communities than to scientists, administrators and political leaders looking at the problem from above. Moreover, not only will local-level priority concerns differ greatly from one situation to another, but they will diverge among social groups. None of the affected groups whether highly vulnerable or not, however, is likely to place land degradation *per se* among their most urgent preoccupations. These immediate concerns usually centre on more immediate survival and livelihood issues for members of vulnerable groups. Maintenance of status and income are usually the key issues for those who are well off.

One should note that to discover how local groups perceive the issues requires skilful in-depth participatory research. Superficial questionnaires about peoples' perceptions are seldom very useful for revealing deeply felt fears, concerns and priorities. Members of poor oppressed groups are frequently unable to articulate their grievances, or more often they may prudently refrain from doing so.

Depending on the context, members of the most vulnerable groups may view day-to-day survival from the ravages of drought or war as the number-one issue. Where their situation is less catastrophic, secure access to land and a little capital such as seeds, tools and animals, or access to some kind of secure employment may take priority. Escaping excessive demands of landlords, moneylenders or tax collectors might also be high on their list, as would be providing some education and health services for their children. Land degradation may be perceived as being primarily caused by drought. Drought is often believed to be an act of God, although local residents may also link it with deforestation or shortened rotations.

The loss of traditional access to land appropriated by outsiders may be regarded as a central issue in some contexts, while in others, where it has been equally serious in its social consequences, it may, like drought, be perceived as being something no one can do anything about at local levels. Prices of the goods and services these groups have to purchase, and for what they have to sell, is nearly always a concern, but again markets and changing price relationships often appear to be beyond any kind of influence on their part. Large producers and merchants, on the other hand, even in poor regions of poor countries, may actively intervene to manipulate local markets and to use political connections to influence public policies.

This attempt to sketch some of the priority issues as viewed from below is necessarily only a caricature. It would diverge in each context and among social groups. It seems likely that rather strong regional patterns would emerge based on shared cultural and historical experiences. Land-tenure issues, for example, seem more likely to be perceived as crucial by the rural poor in much of Latin America and Asia than they are in Sub-Saharan Africa, although the negative impacts of land alienation for the environment and for livelihoods may be similar (Barraclough, 1991a). Vulnerable groups in high-income industrial countries would have their own set of priorities. The concerns just mentioned, however, would be seen by local rural people as issues in one way or another in a great many localities experiencing land degradation. They will be explored a little further in the rest of this chapter.

Grassroots responses

Faced with falling land productivity, often accompanied by increasingly restricted access to land, frequent severe droughts, unfavourable terms of trade, a less than supportive state, and sometimes with devastating armed conflicts, vulnerable social groups in dryland areas are faced with unattractive alternatives. They can try to adapt their traditional production and consumption patterns to cope with these unfavourable conditions. They can attempt to find new sources of livelihood in or near their old communities. They can migrate temporarily or permanently. And they can organize collectively to resist the forces threatening their livelihoods.

Vulnerable pastoralists and peasants frequently resort to a combination of these strategies, but principally they adapt their customary production and consumption patterns, and they migrate. If the literature reviewed is any indication, finding new sources of income locally and organized collective resistance have been successful only in special circumstances. Destitute rural people in subordinate social positions do not have much time, energy or possibilities for taking new organized collective initiatives. If they do take them, they are likely to be ruthlessly suppressed. Here again, there are wide divergences among regions, countries and localities at any particular time.

Adapting production and consumption patterns

The most common strategy is to adjust traditional production and consumption patterns to cope with adverse circumstances. Coping may take place at the expense of the nutrition and health of vulnerable groups as well as at the sacrifice of their scanty assets. This happened on a large scale during the 1990/91 drought in northern Sudan (Buchanan-Smith and Petty, 1992). The IIED/UNRISD/UNEP annotated bibliography reviews a wide range of studies in dryland regions of how vulnerable populations cope with adversity. It distinguishes between traditional coping strategies and more recent ones that have arisen after customary systems were weakened or disrupted by colonialism, modernization, commercialization, political conflict and the like. It also distinguished between strategies adopted by pastoralists and those by peasant farmers. Most of the studies were in Africa but there are also materials from Asia and Latin America.

Pastoralists in the Sahel, for example, had developed complex social structures and sophisticated land management systems to facilitate the minimization of risks associated with drought (Stiles, 1992; Behnke and Scoones, 1992; Speirs and Olsen, 1992). Such strategies were based on mobility, flexibility, diversification and reciprocity.

Nomadism embodies mobility. Nomads move their flocks over wide areas to take maximum advantage of seasonal and cyclical availability of pasture and water (Bonfiglioli, 1992). They also diversify their herds and maintain what many outside 'experts' perceive as an excessive number of low-quality animals in relation to the offtake of meat, milk and other products. When examined in more detail, this practice makes economic sense for the nomads (Frankie and Chasin, 1980; Stiles, 1983; Behnke and Scoones, 1992; Horowitz, 1990). It enables them to make greater use of abundant forage in good seasons and years while reducing the risk of losing nearly all their subsistence and breeding stock during poor ones. Moreover, it permits production of a wide variety of animal products required for self-provisioning (Ibrahim and Ruppert, 1991; Swift, 1973). For similar reasons, they combine animal raising with risky dryland cropping and, when possible, with a little crop production in riverine flood plains where it is less risky. They also develop complementary barter and grazing rights relationships with sedentary agriculturalists as was illustrated in the Algerian case earlier.

Social structures developed to facilitate this risk aversion strategy

Marriage customs, inheritance patterns, inter- and intra-family divisions of labour and property rights had many variations among different groups of pastoralists. All evolved in ways to minimize risks from drought and other adversities. They also institutionalized reciprocal obligations ensuring that risks were widely shared. Common property regimes tended to be the rule in the control of land. Access rights and obligations were usually clear and often very elaborate for different users. Access rights to certain rangeland areas and water holes were limited to particular groups and seasons, for example, while some were reserved exclusively for use during severe droughts (Frankie and Chasin, 1980). Traditional pastoralists in dry regions of Asia and Latin America developed social systems and risk aversion strategies similar in many ways to those found in Africa.

Traditional communities of settled agriculturalists commonly developed strategies in dry regions to minimize risks from climatic variability. These were also reflected in their social structures. Common property regimes for land tended to be the rule. Where intensive agriculture evolved, however, the rights and obligations of each family and individual were clearly defined (Watts, 1987). Otherwise, maintenance of complex canal systems, terraces and soil-enhancing practices, for example, would have been impossible. Settled farmers also benefited from complementary relationships with nomads, bartering their products and benefiting from seasonal depositing of manure.

Traditional pastoral and farming societies have been badly disrupted everywhere by incorporation into colonial, national and global economic and political systems. Their flexibility and other risk aversion strategies have been severely curtailed. Land alienation, together with commercialization of exchange of goods and services, including land and labour, have been accompanied by increasing social stratification and the breakdown of traditional social relations (Dahl and Hjort, 1979; Johnson and Amaah, 1974; Lashova, 1989; Little, 1985; Monod, 1975; Salih, 1985). Traditional social controls that had made these systems sustainable no longer functioned. This resulted in a large proportion of these populations becoming destitute with inadequate access to land and other resources to maintain customary livelihoods or even to sufficient food for health and physical well-being. Disruption of traditional societies has also been an important factor contributing both to rapid demographic growth and massive out-migration from many dryland regions.

One response to the changing context has been to attempt to maintain production using traditional practices to the extent possible and introducing new ones such as chemical fertilizers (if they can be obtained), more cash cropping at the expense of self-provisioning and shortening rotations. New practices requiring purchased inputs, however, usually depend on outside support by governments or others such as agro-exporters. At the same time, consumption patterns are changed to use more purchased products and substitute scarcer local ones such as wood for fuel and construction with inferior ones such as dung and mud-bricks (Barraclough and Ghimire, forthcoming). Obviously, these responses frequently lead to accelerated land degradation and also to social deprivation. On the other hand, in some situations, responses by peasants and pastoralists have been conducive to the adoption of agricultural systems using both land and labour more intensively in ways that could be sustained for indefinite periods (Mascarenhas, 1993; Soussans *et al.*, 1993; Utting, 1993).

The issue is not why some peasant groups are tradition bound, lazy and ignorant while others are hard-working and innovative, as if these were somehow individual or group and community characteristics. It is rather to understand the historical processes which led to sustainable production systems in some circumstances while not in others. This requires analysis of the broader institutional and policy contexts that induce and perpetuate unsustainable responses or encourage more sustainable ones.

Finding alternative livelihoods locally

The option of members of vulnerable groups finding other productive activities locally has seldom been feasible in depressed dryland areas. Traditional crafts and trades are as negatively affected by a shrinking resource base and the inflow of commercially sold cheap manufactured products as are traditional farming and grazing systems. In times of drought, vulnerable groups may find relief locally from food hand-outs by government or international aid programmes. Some may resort to petty trade or menial services. The usual alternative chosen, however, is to migrate.

There are notable exceptions. Where national policies and institutions have been conducive to the development of new rural industries, such as they were in Israel, Taiwan and parts of China after the 1950s, these have occasionally absorbed significant numbers of poor rural people in new lines of production, in constructing complementary infrastructure and in increased economic activity generated by both (Barraclough, 1991a). New investments, however, are always limited by macro-constraints of markets, costs, finance, development strategies and political priorities. Vulnerable groups in dryland regions are usually less well placed than are many others to benefit from such initiatives.

Another exception is seen by the ingenious ways that poor peasants have responded to opportunities to improve their livelihoods offered by new lines of agricultural production and trade in a few sub-humid areas where new cash crops have a profitable outlet. Unfortunately, these are often illegal. For example, where coca, poppy or cannabis production, processing and smuggling for the international drug market has become lucrative, peasants in some sub-humid regions, and in several more humid ones, have found new sources of income near their traditional communities. The reason for citing this anomaly is merely to illustrate that the principal reason most poor peasants do not improve their incomes using their own resources is not laziness, ignorance, lack of initiative or ingenuity, but simply because they do not have a real opportunity to do so.[4]

Migration

Temporary or permanent migration is the usual response when livelihoods can no longer be maintained in the face of land degradation and other factors. In semi-arid areas of Burkina Faso and Senegal, the major sources of income for many communities are remittances from migrants to export crop-producing areas in Côte d'Ivoire, to the cities or abroad, and pensions paid to local residents who once served in the French colonial army (Savané, 1992). Numerous studies in dryland communities in Mexico, Central America, India and Nepal show a similar pattern. Remittances sent by workers in the mines of South Africa help keep alive countless communities in all of southern Africa. These are mostly in dryland regions that have badly degraded lands. This has been well documented for Botswana, although the sources of remittances are becoming increasingly domestic in this relatively prosperous Sub-Saharan country (Buchanan-Smith, 1990).

Recent migrants from Central America to the United States mostly came originally from degraded dryland regions of Guatemala, El Salvador and Honduras. They often left after a stay of some years, or a generation, in these countries' cities where they received some basic formal education. The value of remittances in these Central American countries by legal and illegal migrants was estimated in 1990 to be greater than the total value of their traditional agro-exports (CEPAL, 1992).

In Brazil, a large portion of the urban workforce of the industrial south, and also of workers and colonists in Amazonia, migrated from the dry north-east. For reasons explained earlier, such migrants should not be considered to be primarily environmental refugees even though a high proportion left in drought years. Many other factors such as an exploitive land tenure system, government policies, population growth, terms of trade, the search for new opportunities in the cities and the like would have resulted in large-scale migration even without land degradation and droughts. But the latter were contributing factors.

Collective resistance

The last option of vulnerable groups is organized collective resistance in the face of threats to their livelihoods posed by land alienation for development projects, for agro-export production by large operators, and other uses of land such as for mines or protected areas. Efforts to resist have been numerous, but seldom successful. To be effective, such resistance has to find powerful allies in the broader society. Most collective initiatives to resist land alienation are crushed by local authorities and elites without ever attracting attention nationally or internationally. A few, however, find allies and may achieve some of their objectives, at least temporarily.

A good illustration of the former was the attempt during the last two decades of Guatemalan indigenous groups to resist appropriation of their traditional lands in Altaverapaz and several other regions. They were then accused of supporting the guerrillas and many were massacred by the army. Those who escaped fled to seek survival by cutting and burning forests to plant subsistence crops on steep mountainous terrain that was very difficult for the army to penetrate (AVANCSO, 1992; and personal interviews with refugees).

The conflict between the semi-nomadic Barabaig and the Tanzanian National Agriculture and Food Corporation (NAFCO) illustrates limited success. With the technical and financial support of the Canadian development agency CIDA, NAFCO is converting over 100 000 hectares of their traditional grazing lands into wheat production. The Barabaig attempted to resist through legal and political channels. There was some limited violence and several human rights violations. The Barabaig found allies who could help them plead their case both nationally and in Canada. The wheat project is very dubious from an economic standpoint in any case. The Tanzanian state was less authoritarian and more respectful of legal procedure and human rights than was that of Guatemala. The conflict is not yet resolved, but the Barabaig's collective resistance has probably prevented more of their lands being taken with no compensation. It may lead to changes of some policies both nationally and by CIDA. It could possibly stimulate more realistic development projects in dryland regions both in Tanzania and elsewhere in the future (Lane, 1990, 1993).

Collective organization by the victims is essential to resist the negative impact of 'development'. Otherwise political authorities and other elites, including foreign aid agencies, are likely to ignore them. If they view poor vulnerable groups as potential allies or as opponents who can cause trouble, however, their grievances may be taken more seriously. Collective resistance can be very risky for the powerless. Much depends on the nature of civil society and of the state in each situation.

Social impacts of policies and programmes to control desertification

Many colonial administrators in the 1920s and 1930s in dryland Africa became alarmed at the mounting evidence of accelerating soil erosion, deforestation and rangeland deterioration. They usually associated these processes with peasant ignorance and short-sightedness. Population increase only began to be seen as a serious problem in most of these colonies much later, although population densities often had increased alarmingly in areas reserved for traditional agriculture due primarily to alienation of good land for white settlers or its diversion from traditional self-provisioning uses to cash crop production. Colonial authorities were frequently concerned about labour shortages for colonial enterprises. In fact, populations actually had decreased in many dry regions under the early impact of social disruption accompanying colonial conquest.

Colonial conservation programmes

Colonial governments concerned with land degradation often attempted to regulate land use in accordance with conservation norms that had developed in Western Europe and the United States under very different social and ecological conditions (Beinart, 1984; Cliffe, 1988). For example, 'natives' in some areas were compelled to dedicate several days of labour each year to constructing bunds or terraces intended to control soil erosion on hillsides. They were forbidden to exploit many forest areas for fuel, construction materials or fodder that they had been using rather sustainably for centuries. Traditional use of fire in land management was often proscribed (IFAD, 1992; Blaikie and Brookfield, 1987; Gadgil and Guha, 1992). Also, large areas were withdrawn from traditional uses to become parks and game reserves.

Naturally, local people came to associate conservation policies with oppression. This was even more the case, because the conservation measures were frequently counter-productive. These conservation policies were not based on an understanding of local social systems, farming practices or ecologies. Only a few sophisticated colonial observers linked desertification to social disruption and land alienation instead of short-sighted exploitation by native pastoralists and cultivators.

This colonial syndrome in Africa merely repeated with variations what had happened earlier in India, the Americas and many other places. Governing authorities tended to blame the victims of desertification for its occurrence. New elites, and not a few older ones, were happy to do the same. Conservation programmes were designed to attack the symptoms, but not the real causes of desertification. To a great extent, desertification control programmes continue to do the same nearly a century later, although now national governments and other outsiders instead of colonial authorities are sponsoring them.

Recent programmes in poor countries

In recent years, national governments, international and bilateral agencies as well as many NGOs have launched programmes and projects in developing countries to control desertification. The literature suggests that these have not been very effective on the whole. The desertification problem remains about the same as it was in the 1970s, if not worse (Grainger, 1988). 'Successful' projects tend to have been small and of questionable replicability. Moreover, there is considerable disagreement about what constitutes success. Land conservation objectives may be achieved, often at considerable financial costs met by foreign donors. Nonetheless, this may be done at the expense of the rights and livelihoods of many traditional users of the land.

The Tanzanian government's HADO project is an example. It was funded by Swedish SIDA in the semi-arid Kondoa region of over a million hectares, one tenth of it badly eroded. The project began in 1973. In 1979, traditional pastoralists with their animals were completely removed from the degraded area to permit vegetation to recover aided by various other rehabilitation measures. One source cites it as an outstanding success. It concludes, 'The rehabilitation of the vegetation in this area, already considered irretrievable in the 1930s, has been remarkable and more cultivable land has become available for farmers' (IFAD, 1992). In contrast, a Tanzanian analyst, after considerable long-term field research in the area, considered it something of a failure. He concluded 'The eviction of people and cattle can restore vegetative cover in a relatively short time. But these sectoral and technocratic approaches only export the problems to other parts of the district, region and even beyond . . .' He adds that the project was extremely costly in money and trained staff so that there would be little possibility of replicating it in other parts of Tanzania (Mascarenhas, 1993).

Even if there is some agreement on success and failure criteria, there are still the problems of time and contexts. A project that appears to be successful after five years may look like a disaster after twenty years, while one that looked like a flop after five years may be regarded as a success after twenty (Nelson, 1988). The dynamics of each situation is different. In part this is due to changing national and international contexts that are practically impossible to take into account adequately. Moreover, these interact with highly variable local-level factors such as leadership, participation, technical skills and resources.

Programmes and projects intended to control desertification have usually been designed to attack its direct causes as perceived by their administrators or donors. Rangeland rehabilitation schemes such as the Tanzanian one mentioned above and many other attempts to introduce 'improved' pasture and livestock management practices have been common. So too have the construction of boreholes to provide water for pastoralists' animals in dry regions. Results have often been disappointing. Evidence has been accumulating that traditional nomadic and semi-nomadic pastoralists using opportunistic grazing strategies can frequently make more economic and sustainable use of these low-yielding dry rangelands than can be done by modern ranching methods with higher-quality animals and 'optimum' stocking (Behnke and Scoones, 1992). Boreholes, by inducing a concentration of animals around them, often lead to accelerated land degradation (Frankie and Chasin, 1980), although this is not always the case (Nelson, 1990).

The dilemma facing pastoralists is that in most places they no longer have access to much of their traditional territories. Moreover, their complementary relationships with sedentary neighbours have usually been largely severed and they themselves have become much more dependent on earning cash for market purchases while at the same time becoming more socially stratified, as was seen earlier. Not only are they unable to expand their herds more than they have in most situations, but new settlers, development projects, protected areas and the like continue to encroach their traditional territories. In some areas shortages of fuel from brush and trees may be as severe a constraint as is scarce pasture and water.

Not surprisingly, planners and administrators, viewing these same problems from a quite different angle than the pastoralists, often conclude that the only way out of the dilemma is for nomadic pastoralists to become sedentary crop producers, tree farmers and livestock growers. This implies programmes to stimulate more intensive dryland cropping, tree planting and, where possible, irrigation, accompanied by some animal husbandry. A great deal of effort has been made along these lines. These new farming systems often yield meagre or negative economic returns while being unsustainable ecologically and socially. The Tanzanian wheat project mentioned earlier was an example. Drought-relief programmes in higher-income Botswana were relatively successful in transferring some income to poor rural residents, but those aimed at increasing agricultural productivity primarily benefited well-off farmers and ranchers (Buchanan-Smith, 1990).

Examples of irrigation and settlement schemes

Irrigation and resettlement schemes have been a common government response to rural poverty and land degradation problems. In Brazil's north-east, for example, substantial public investments were made during the last few decades in the construction of several large publicly controlled reservoirs, hundreds of smaller private ones, and for building irrigation systems. These were often accompanied by costly small farmer resettlement projects.

Irrigation projects absorbed nearly half the region's rural development budget during the 1970s and 1980s. Results were disappointing for several reasons. The large dams were built without taking into account the complex hydrology of the Sertâo. In consequence, the new reservoirs often affected negatively other traditional water resources. Also, some irrigated cropland soon became too salty to use. Costs were excessively high per hectare and per person benefited. The region's extremely skewed land tenure system and social structure made it inevitable that large landholders would benefit most, especially as they were the best placed to use highly subsidized irrigation water for the cash crops the government was trying to promote. Moreover, most of the smaller private reservoirs were located on large holders' properties. Much valuable crop and pasture land was inundated by the large reservoirs and many smallholders were displaced with little or no compensation. Few small farmers were actually resettled.

This costly programme made no dent in the north-east's widespread rural poverty nor did it slow land degradation or significantly increase agricultural output. It enjoyed considerable political support, however, because many influential large landowners, contractors, politicians and public officials derived substantial private benefits from it, even if most peasants did not (Hall, 1983; Droulers, 1989; Cavalcanti, 1989; Livingstone and Assunçâo, 1989).

The social impact of this ambitious irrigation programme was similar to those of countless other land development projects in Brazil. The rapid expansion of mostly unsustainable low-quality cattle pasture in Amazonia at the expense of tropical forests and their traditional users is another example (Diegues, 1992; Barraclough and Ghimire, forthcoming). The principal explanation lies not so much in 'faulty' project design or even corruption as in public policies. These in turn were closely linked to the country's underlying socio-economic structure and its interface with the world system.

Irrigation and other rural development projects in dryland regions with many similar characteristics and social results have been described in the Sudan, Senegal, India and several other countries. The same is true of afforestation projects, farm forestry programmes and initiatives aimed at improving social and economic infrastructure. Only a few achieve their stated socio-economic and ecological goals. Far fewer can be regarded as being sustainable or replicable.

Popular participation

Part of the literature emphasizes that successful projects and programmes should be sensitive to the self-perceived needs and aspirations of the vulnerable groups they are supposedly intended to benefit (e.g. Stiles, 1987). There is a growing consensus that these groups should be active participants in programme and project design, administration and execution. Many studies insist on the virtues of small decentralized projects and programmes in contrast to large centralized ones. Several authors warn against top-down technocratic solutions. A few call attention to the need to take into account the socio-economic and political dimensions of desertification at all levels in designing programmes to control it. Global reviews of desertification programmes repeat again and again that there should be better co-ordination among government and international agencies, NGOs and donors to permit an effective integrated approach to desertification problems regionally, nationally and at local levels. But the literature reviewed is not very helpful in suggesting how these laudable goals can be met in diverse specific contexts.

This large gap in the literature is easily understandable. One is dealing with fundamental issues of social relations, political power, economic growth and social reproduction that have plagued societies throughout history. The social issues of desertification are, like the ecological ones, to some extent locality-specific, but they are also to a very important extent national and global issues that cannot be dealt with effectively locally without supportive initiatives nationally and internationally. As was seen above, national and international markets, policies and political relations have been the principal driving forces behind the processes leading to land degradation in dry regions as well as to the increasing social polarization of their societies. All these factors contribute to widespread poverty and high levels of social conflict.

Key societal issues

This final sub-section briefly looks at wider social issues. These embody many of the principal constraints and opportunities for dealing constructively with the social causes and implications of desertification. They are issues that have always been part of the development debate. The difference is that now there is growing recognition that society at the global level may already be approaching the environmental limits of economic and demographic growth in their strictly quantitative sense. Adjustments at local levels to encourage higher productivity and more adequate living levels of vulnerable poor groups may increasingly require compensatory adjustments in the composition of production and consumption, and of technologies, of higher-income groups elsewhere.

If 'development' is going to become sustainable globally, economic growth will gradually have to become qualitatively and quantitatively different in practice from what it has been in the past. Perhaps dealing effectively with the social issues of desertification in several localities can help societies to grope haltingly towards more sustainable development. To do this, the wider implications of local initiatives have to be explicitly taken into account.

The preceding discussion emphasized that the social issues of desertification are always to some extent locality-specific. This is why generalizations frequently tend to be rather banal. The discussion also brought out that the principal socio-economic processes and institutions stimulating land degradation do not originate locally. Remedial initiatives have to be national and international as well as local. Generalizations about societal constraints and opportunities, however, are as messy as they are about local ones.

Four clusters of national and international socio-economic and political issues appear to be especially pertinent for improving natural resource management in dryland regions. The first includes the public policies and market forces that are usually considered central for any development strategy. The second concerns institutional structures and related policies regulating land tenure, land use and social relations more generally. The third deals with farming systems, technologies and broader economic structures that constrain alternatives for vulnerable groups and sometimes offer new opportunities. The fourth set includes demographic trends and their implications. A better understanding of how vulnerable groups and other social actors with many conflicting interests perceive these issues is a key ingredient for dealing with them constructively.

Public policies and market forces

The policy context was crucial for understanding the social dimensions and dynamics of land degradation in all the cases reviewed. Colonial and national trade, price, tax and investment policies drove the introduction of cash crops, often at the expense of food production locally. These policies led to large-scale alienation of lands previously accessible to vulnerable groups and to the new social stratification accompanying commercialization of traditional economies. Social policies or their absence largely determined the access of different rural groups to 'modern' health and educational services. Recently, the relatively low priority given to social services in tight budgets by both national states and international financial institutions have led to the curtailment of the inadequate social services that may have existed in many dry regions. Investment policies frequently accelerated the marginalization of vulnerable groups, the plundering of their traditional resources and the failure to create alternative sources of livelihood. The constant wars plaguing the Sudan and many other dryland regions were the biggest single cause of famine. Such conflicts can be considered a result of policy failures both nationally and internationally.

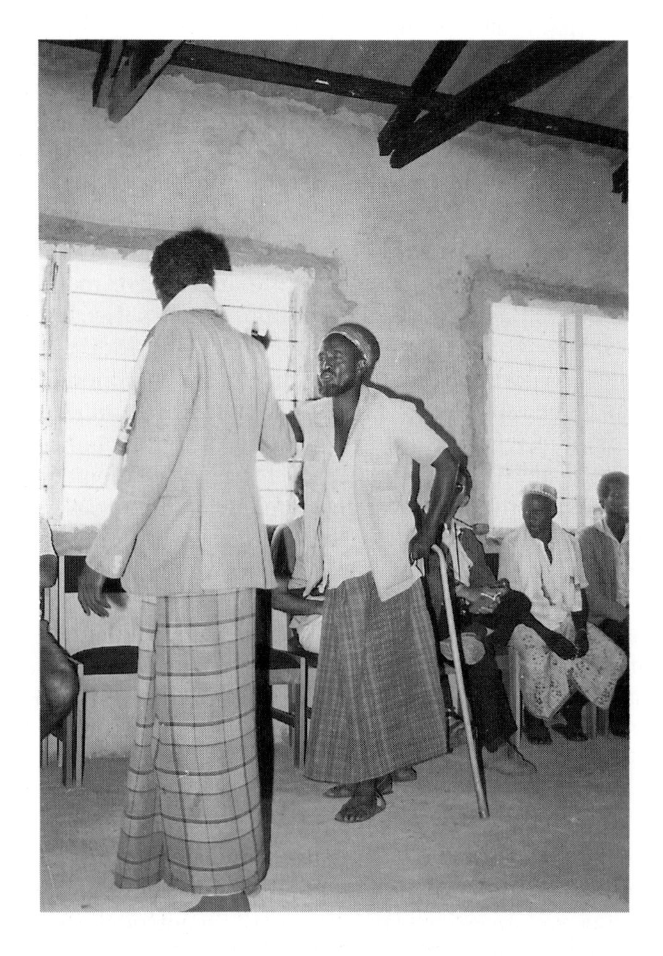

Photograph 2.4. There is a growing realization that projects will only be successful if local people have a say in their design and implementation. This Somali man is presenting his views on what needs to be done in northern Kenya at a UNEP-sponsored pastoralists' workshop. (Photo: Daniel Stiles)

Numerous analysts have argued that a prerequisite for any kind of really sustainable development is a popularly based development strategy. Trade, price and credit policies, social policies and investment policies should be a part of a broader strategy aimed at meeting the basic needs of vulnerable groups and creating opportunities for them to improve their livelihoods both qualitatively and in terms of food security, living levels and the like (Barraclough, 1991a). Such a strategy, however, implies popular participation. There have to be local institutions and some kind of a national state or similar political entity in which these groups are strongly represented and that is accountable to them as well as to other hitherto much more powerful interests. It also implies a more democratic and accountable world order with some sort of effective system of governance capable of resolving international conflicts and guiding global economic and political processes towards more sustainable goals. The institutions and policies required to control desertification are somewhat utopian. This does not preclude partially effective local and national initiatives in their absence, but it greatly circumscribes the possibilities. This is especially the case in places where vulnerable groups are deprived of their civil and human rights or where armed conflicts are dominant.

In most places there is some limited scope for reforming policies so that they provide at least some vulnerable groups with better livelihoods as well as greater possibilities and incentives to use land more sustainably. For example, emergency aid to drought victims and refugees could more frequently be made part of longer-term efforts designed to upgrade both human and natural resources. Food aid and cheap subsidized food imports do not necessarily have to be administered and used in ways that contribute to depressing the incentives of local farmers to increase their own production, and to diminish the urgency for governments to adopt peasant-based rural development policies. Political support can frequently be found or generated to give a higher priority to enabling vulnerable rural groups to have better access to the social services they deem important and for increasing social investments. Public and private investments in infrastructure, technologies and productive capacity could be guided more directly towards employment creation and meeting popular needs. Social costs associated with land degradation, such as accelerated migration, siltation, pollution, reduced biodiversity, a depleted water reserve and the like could be explicitly recognized in public and private accounting systems, supplemented by taxes and standards that induce those who are ultimately responsible, and able to pay, to change their ways or meet the costs (Reed, 1992). International monetary and trading systems could be reformed to be more equitable and supportive of sustainable popularly based development strategies (Daly and Goodland, 1993). In this respect, easing the pressures on developing countries to expand their exports by overexploiting drylands and other natural resources in order to service odious foreign debts could be particularly helpful (Adams, 1991).

But for these favourable policy outcomes to occur, there has to be 'political will'. This means much more than convincing a few political leaders and planners. It usually implies, among other things, the autonomous organization of vulnerable social groups and a political system that induces the state to be increasingly responsive to their legitimate demands. It also implies an increased capacity of developing countries to work together in international organizations and fora (South Commission, 1990).

Land tenure

Land tenure issues were prominent in nearly all the literature reviewed. The significance of land tenure, however, was frequently underestimated, while the difficulties of reforming it were often exaggerated or merely wished away. Property relations are fundamental in determining what groups have access to resources on what terms and who gains or loses during 'modernization' processes.

Property is essentially a sub-set of social relationships sanctioned by law and custom. Land tenure implies a bundle of institutionalized rights and responsibilities. Land tenure institutions regulate relationships among individuals, families, social groups and classes, corporate entities and the state in their access to land and its products, including the rights to anticipated future benefits. Clear and secure tenure rights and obligations are a necessary but not a sufficient condition for individuals or groups using the land to invest labour and other scarce resources in conserving or improving it. The benefits from such investments can not usually be realized until some future date.

The necessary conditions of clear and secure tenure rights can be realized under a variety of land tenure regimes whether they be ones of private, communal or of state ownership. When land tenure rules are capricious or break down under outside pressures, however, the tenure regime tends to become one of open access. This can happen whether the tenure regime is formally one of private, state or communal ownership. An open-access regime implies that everyone's property is no one's property and often may lead to the 'tragedy of the commons' as a result of many competing users attempting to mine all the short-term gains possible before others can do so first. This, of course, directly stimulates land degradation. State, private and communal tenure regimes can all be conducive to land degradation even when rules are apparently clear and tenure secure, if the only way for those depending on the land to survive is to overexploit it. The same occurs if the broader social system makes it lucrative for a wealthy land manager to abuse the land and invest the gains elsewhere in other activities.

In many dry regions, land tenure is extremely insecure. Customary communal tenure rules may still reign locally, but the state may have adopted other norms of private or state ownership systems that can be arbitrarily imposed at any time when it is advantageous for state authorities or some of the state's powerful support groups, such as large agro-exporters, hydroelectric developers, tourist interests or national and transnational timber enterprises. Moreover, most pastoralist groups have already lost access to much of their customary territories. Also, traditional agriculturalists have often been forced to shorten their rotations drastically and to concentrate increasingly on cash crops. Where peasants and pastoralists have successfully intensified production, insecurity generated by exploitive marketing systems and corrupt officials can lead to non-sustainable land-use practices just as surely as can insecure land tenure. In many dry regions the best agricultural land has already been appropriated by a relatively few large producers or landlords.

Insecure land tenure and inadequate livelihoods induce poor land users who can barely survive to mine their resources unsustainably. Calling this a time preference of the poor for immediate over future benefits, as some analysts have done (IFAD, 1986; World Bank, 1991), seems a little bizarre. Pastoralist and peasant groups even when desperately poor often managed their land sustainably for centuries. Through trial and error they learned that the optimum rate to exploit a renewable resource was one that would maximize its sustained yield irrespective of the implicit discount rate (Dasgupta, 1982). This was in large part because they did not perceive viable options for survival from other sources. On the contrary, private capitalist or state land managers may in some circumstances deliberately overexploit their resource because they have the option of reinvesting the profits in other much higher yielding activities while passing on the costs of desertification to the rest of society and to future generations.

Reforming land tenure systems is always difficult because it challenges established social relations more generally. Reforms that provide better and more secure access to land by vulnerable groups usually imply that powerful local elites or outsiders have to give up some of their privileges and claims to future gains. The latter may be largely speculative, but meanwhile many landless and near-landless are denied secure access.

Land reform of some kind is essential in most dryland contexts if desertification is to be controlled at acceptable social costs. This could imply providing the state's legal sanction and protection for customary land systems in many circumstances. The economics of managing low productivity range land and many forest areas are such that efficient and equitable common property regimes that are self-administered would often be preferable to the expensive cadastrals and other legal-administrative costs implied by effective private or state ownership. In many agricultural regions where customary tenure has already broken down, and especially where good land has been monopolized by a few large holders, redistributive reforms may be required dividing the land among individual families, or possibly small co-operatives. The political obstacles are always great.

Depending on the context, the goals of land reform can sometimes be approached through credit and tax reforms, complemented by ecologically based land use zoning. Access to credit in favourable terms for land acquisition and production by the landless and near-landless could help these vulnerable groups take advantage of their underused labour and managerial potentials. Progressive land taxes that exempt smallholders with access only to what is required for subsistence but that charge taxes approaching rental rates for the land held by speculators and large commercial producers could also theoretically be very effective tools for bringing down the price of land for the poor while stimulating improved land use. They could penalize unsustainable uses by helping to internalize the costs of careless land management.

Such an approach, however, is politically and administratively difficult. It is usually impossible in poor countries and regions. To be feasible, it would have to be locally administered by the communities affected with the support of state power when required. Both communities and the state would have to be sufficiently democratic so that the credit, tax and zoning rules would benefit important groups of the rural poor and provide them with real security and incentive for sustainable land management. Landholding elites are usually very powerful, enabling them to use credit and tax reforms ostensibly designed to benefit the rural poor and protect the land to strengthen their own privileged positions (Barraclough, 1991a). In spite of these difficulties, land reforms are essential for improved natural resource management in most dryland regions. More effort has to be dedicated to bringing them about.

Farming systems

Developing more productive and sustainable farming systems should have a high priority in dryland regions. Such farming systems could contribute to improving the food security of vulnerable groups while at the same time protecting land resources from degradation. This issue, of course, is closely related to those of policies, markets, land tenure and demographic change. Moreover, just as land tenure issues imply those of social relations more generally and policy issues suggest those of political power, governance and the nature of civil society and of the state, farming system issues directly call into question much wider ones of technology and economic structures locally, nationally and globally.

The literature review strongly supports the notion that traditional cropping and pastoral systems, if judged by environmental and livelihood criteria, have often been superior to modern ones depending on new technologies and the purchase of many externally produced inputs. Indigenous systems were often less risky, more equitable and made fuller productive use of available human and natural resources. Modern science and technology can make a big contribution. Nonetheless, it is much more difficult and complex than has been widely assumed to introduce modern technologies without their being accompanied by negative social and environmental impacts. To the extent possible, improved farming systems should be based on the accumulated knowledge and experience of indigenous populations.

Low-external-input systems tend to be advantageous also because they are less disruptive of traditional social systems and minimize dependency of local people on volatile terms of trade in national and international markets. Especially in poor countries with small industrial capacities, external inputs of chemical fertilizers, pesticides, machinery and the like, as well as of petroleum, have to be imported from abroad. This increases national dependency on aid, exports and loans. Small farmers and pastoralists dependent on a high proportion of externally purchased inputs are also often exploited by intermediaries and officials at all levels. Moreover, high external input farming systems tend to be intensive in the use of energy from fossil fuels both directly and indirectly while economizing in the use of labour, which is often the low-income farming community's most abundant resource.

One should be realistic, however. Where opportunities exist for financially profitable lines of production for national and international markets that require purchased inputs, outside entrepreneurs are probably going to exploit them sooner or later with or without the consent and co-operation of local groups. Moreover, many traditional or transitional farmers and herdsmen will want to enjoy the conveniences and perceived benefits of labour-saving machinery, chemical inputs, manufactured consumer goods, including eventually such 'luxuries' as televisions and private cars enjoyed by the rich in their own countries and by their farmer and worker counterparts in the industrialized ones. Traditional farming systems are vulnerable to the triple pressures of the search for profits by outsiders, the consumerist aspirations of the poor still struggling to survive and the drive by state and private 'developers' to 'modernize' 'backward' societies and social groups. An external context of strong and well-designed popularly based policies and institutions is essential for improved indigenously based farming systems to prosper. These problems are frequently made more acute by population growth in rural areas where alternative sources of livelihood are scarce and unattractive either locally or elsewhere.

Another problem is that indigenous farming systems and environmental lore are always rather site-specific. A few aspects can be transferred and diffused elsewhere, but this is the exception that proves the rule. Similarly, improvements or adaptations based on modern science also tend to be difficult to incorporate in diverse ecological contexts. This implies a great deal of decentralized research with specific local groups before possibilities of improving sustainable outputs by incorporating innovations into their farming systems can become a reality. Such research has to be both socio-economic and technical. It requires an understanding of the implications of introducing innovations for local livelihood systems and the natural environment.

Local farming systems everywhere are becoming increasingly influenced by the production and consumption patterns (with the technologies these imply) that are dominant in national societies and in the industrialized countries. These industrial systems dominate national and international markets. They

Photograph 2.5. These people organized tree-planting and nursery activities in India as a community participation project supported by UNEP and the government of India. (Photo: Daniel Stiles)

largely determine what is available commercially, and at what relative prices, in the way of consumer goods, production inputs, capital goods and technologies. This is the case even in remote rural areas. Countries such as China and India that emphasized local self-reliance after 1950 are becoming more and more integrated into a global industrial system (Gadgil and Guha, 1992). This is in part a result of their 'development'. But it will not be sustainable unless ecological and social issues are successfully resolved.

Ultimately, the issue of developing and maintaining sustainable and equitable farming systems in poor dryland regions is closely linked with national and global economic structures. These will eventually have to be reformed in practice towards more sustainable, less polarizing production and consumption patterns. Meanwhile, much can still be done in many places to improve farming systems and other sources of livelihood locally, sub-nationally and nationally. But it will be inevitably partial, transitory and extremely difficult.

Demographic change

Continued rapid population growth in most developing countries for the coming several decades is a reality that cannot be wished away. Although the rate of increase is slowing in several countries, it still implies a doubling of the world's population during the next four decades. Most of this increase will take place in developing countries in spite of a trickle of emigration to a few of the rich ones.

The current stress on the global environment, however, is primarily the result of production and consumption in the rich countries that are responsible for some four-fifths of energy consumption and industrial pollutants. The poor consume very little in comparison in spite of being the vast majority. The so-called population and environmental crises are rooted in environmentally and socially unsustainable socio-economic structures and consumption patterns benefiting primarily relatively small minorities of the world's population. The social and environmental implications could be truly catastrophic if past trends were to continue, or if they were to change only to the extent that technologies become marginally less wasteful and polluting while an ever-growing number of people in poor countries adopt today's rich country life styles. A goal of global population stabilization towards the end of the next century and zero net growth thereafter is the only realistic one for a sustainable planet according to many analysts who have seriously considered the ecological, social and political implications of a continuation of present demographic and socio-economic trends (Goodland and Daly, 1993). But this implies major modifications in production and consumption patterns of the wealthy as well as improved living levels and opportunities for the poor.

Desertification and deforestation are commonly blamed on population growth and poverty. In one sense, this is a truism because if there were no people there could be no anthropogenic causes of desertification. Numerous case studies, however, suggest that the relationships between population dynamics and environmental degradation are much too complex to support sweeping generalizations about cause and effect (Ghimire, 1993). In some contexts accelerating land degradation at local levels seems to be closely associated with rural population increase. Closer examination, however, usually reveals that land tenure systems, the expansion of commercial farming or ranching and various public policies were much more convincing explanations of desertification. In some cases, population increase stimulated the adoption of more sustainable farming systems and land-use practices than had prevailed when population densities in the same region were considerably less. In others, as was seen earlier, declining rural populations due to rapid out-migration, war or epidemics were accompanied by accelerated land degradation. There were simply neither the incentives nor enough labour in these depopulated communities to maintain traditional terraces, canals and land protecting practices once the population dropped (Barraclough and Ghimire, forthcoming).

Population projections for the next few decades at national and international levels have to be taken very seriously in planning development strategies and desertification control programmes. Locally, however, population growth may often be reversed rather rapidly, or population may unexpectedly explode, due principally to migrations. There are no simple predictable cause-and-effect links between population growth and land degradation at either national or local levels.

Population growth is seldom perceived as an urgent problem by the vulnerable groups affected by droughts and falling land productivity. The style of development generated by the expanding world system, of which they are among the principal victims, creates strong incentives for them to have many children. Large families are often seen as a source of badly needed family labour, as insurance for old age, as potential migrants who could send remittances, and to ensure family reproduction in the face of traditionally high child mortality rates. Slowing population growth is going to require much more than contraceptives and family planning. Among these other factors are improved food security, good education and health services, better security for the aged, more opportunities for women and improved social and economic conditions generally for vulnerable low-income groups. These requirements converge with those that could also contribute to a social context in which improved natural resource management becomes feasible. There are no simple linkages between land degradation and population growth, but real social development would greatly facilitate the control of both.

III. SUMMARY OF TENTATIVE CONCLUSIONS

This selective review of the literature allows several tentative conclusions concerning the social dimensions of desertification. Perhaps they should merely be considered as working hypotheses. These conclusions implicitly suggest recommendations for UNEP and others attempting to improve natural resource management in dryland regions.

Recommendations, however, tend to be mere platitudes if they are not based upon a good understanding of the objectives, constraints and possibilities of the different social actors for whom they are intended and also of those who will be affected if they are implemented. Hence, this concluding section is followed by an appendix pointing to a few additional research priorities.

The concept of desertification

This review of social issues attempted to follow the UNCED definition of desertification as being 'land degradation in arid, semi-arid and sub-humid areas resulting from various factors including climatic variations and human activities'.

An ambiguous concept

Application of this concept in practice becomes very problematic as a guide for analysing the social dimensions of desertification. Land degradation is a social concept that involves judgements about what constitutes production and productivity. These are value judgements that will vary from one social context to another. Even when there is ideological agreement about what criteria to use, there are numerous technical difficulties and controversies in measuring and monitoring dryland degradation. Moreover, degradation results from countless combinations of natural and social processes that differ widely from place to place and from time to time. It also takes place in diverse ecosystems that exhibit divergent capacities to recover following natural disturbances or those caused by humans.

The ambiguities inherent in the concept of desertification make it more useful for mobilizing political support to combat what is often imagined to be advancing deserts than for analysing the multiple natural and social processes generating various kinds of land degradation and the social impacts associated with them. The implications of these processes have to be well understood in order to propose effective remedial actions leading to improved natural resource management in dryland regions.

Use the term desertification prudently

The term desertification should be used prudently in order to avoid cynical disillusionment by governments and donors mobilized to combat it when they discover how imprecise and controversial the concept really is. Nonetheless, they would often be willing to devote resources to support programmes promoting improved natural resource management if they were based on solid diagnoses of the issues.

The extent of desertification

There is a great deal of disagreement about the extent of desertification and the numbers of people at risk. This is in part a result of the conceptual and measurement difficulties mentioned above.

Areas affected

UNEP's and UNCED's estimates of 3.6 billion hectares in dryland regions undergoing land degradation lose credibility when one discovers that soil degradation was believed to have taken place in only about 1.04 billion of these hectares. While many changes in vegetative cover alone are clearly land degrading, the longer-term impacts on land productivity of many other vegetative changes often called degrading are highly debatable. There is a strong case for placing greater emphasis on the extent and severity of different kinds of soil and vegetative changes when discussing the extent of land degradation in dry regions. Lumping them together under the rubric of desertification can be misleading, especially when attempting to assess the social dimensions of land degradation.

Number of people at risk

The estimate of 900 million people at risk from desertification is based on FAO estimates of the agricultural population in dryland regions where land degradation is believed to have occurred or is in danger of occurring. This datum is not very helpful in assessing the social dimensions of dryland degradation. Large social groups both within and outside of dryland regions may be seriously affected to the extent lower dryland productivity may influence global markets and prices or contribute to large out-migrations. Much larger groups may be at risk in the future due to narrowed options associated with eroded soils, diminished biodiversity and local climate changes. There is little hard evidence that dryland degradation has much influence on global climates. Externalities such as increased flooding, lower water tables, sedimentation, fuelwood shortages and the like associated with dryland degradation may affect large urban populations. Finally, different social groups are affected in diverse ways. A few may even reap benefits from land degradation in some circumstances.

More emphasis could usefully be placed on indirect risks from desertification for large population groups due to externalities and narrowed options for future generations. There should also be greater recognition that different groups of present-day agriculturalists and pastoralists are not equally threatened by dryland degradation.

Causes of desertification

The UNCED definition of desertification clearly states that it is a result of various factors including climatic variations and human activities. Diverse combinations of natural and social processes cause dryland degradation of widely varying types and severity in different times and places.

Interacting causes

Historical, archaeological and palaeoecological evidence indicates that desertification in the past was frequently caused by climatic changes that were not influenced by human activities but that nonetheless resulted in devastating social impacts. In other cases, short-sighted or careless management of natural resources were among the principal causes of destructive deforestation, soil erosion, salinization and waterlogging. Currently, there is compelling evidence that human activities may be influencing global climate changes and will probably do so more in the future, principally through greenhouse gas emissions from industrial sources. There is still much uncertainty about this, however. In any event, dryland degradation apparently plays a minor role in global climate changes, although its impact on local climates can be great in some circumstances and minor in others.

Direct versus indirect causes

The so-called direct anthropogenic causes of land degradation are to some extent tautological. Overcultivation, overgrazing, careless irrigation, deforestation and the like are often said to cause land degradation. At the same time, they are frequently used as criteria for estimating its occurrence and severity. This is similar to saying that poverty is caused by low incomes or drought by an absence of rain. Explanations of social causation imply analyses at local, national and global levels of multiple interacting relationships and processes generating dryland degradation and non-sustainable development.

The multiple interacting indirect causes of desertification have to be confronted in order to design effective policies leading to improved natural resource management at local levels. Commonly cited indirect causes such as demographic growth, policies and market forces, consumption patterns, technologies, land tenure systems, climatic change and the like can have divergent and often contradictory impacts in different contexts.

Flexible policy mixes have to be adapted to the unique characteristics of each locality. Simplistic attributions of dryland degradation in general to only a few direct or indirect causes is usually misleading.

Social impacts of desertification

Divergent impact for different groups

Dryland degradation has divergent impacts for various social groups. Social consequences will also diverge in different contexts. In some circumstances drought and land degradation can lead to small minorities benefiting at the expense of more vulnerable groups. Moreover, the impacts for many non-agricultural groups in addition to dryland cultivators and pastoralists can be significant. This is due to externalities such as reduced water supplies, higher food and fuel prices, floods, pollution and accelerated out-migrations, as well as to reduced options for unborn generations.

Importance of externalities and reduced options

Greater attention could usefully be paid by UNEP and other organizations combating desertification to indirect social consequences of land degradation due to externalities and narrowed options affecting dryland agriculturalists and pastoralists as well as much wider populations both now and in the future. At the same time, there should be more recognition of the differential implications for various social groups believed to be at risk because of their direct dependence on degraded drylands.

The remaining conclusions deal in more detail with social causes and consequences of desertification, taking into account the perceptions of vulnerable groups.

Local-level perceptions and dynamics

Local social systems in dryland regions are highly diverse with unique historical and ecological contexts. Those proposing national or international programmes to improve natural resource management should be acutely sensitive to local differences. They have to think globally but act locally. At the same time, they have to act globally and nationally taking into account the wide range of local conditions and perceptions that their policies are designed to influence.

Elusive land managers

Key land managers are very difficult to identify in customary communal systems. Decisions concerning land use and pastoral or farming practices, as well as those about investments of various kinds, marketing and consumption, are often widely diffused within communities and extended families according to complex traditional rules and inheritance patterns. In many latifundia systems with multi-layered subtenancies and debt peonage it is also very difficult to find the key land managers. The same is true of smallholder systems in which cash crops are being financed by or shared with landlords, merchants or moneylenders.

Who makes which decision has to be investigated through anthropological sleuthing in a wide variety of contexts. Without such information, combined with an understanding about the objectives and perceptions of the different decision makers, it is practically impossible to design effective local-level programmes to improve land management. Most poor peasants and pastoralists, for example, are primarily interested in maintaining and improving their meagre livelihoods. Preventing land degradation is usually a concern only to the extent it affects their whole livelihood system. Better-off groups such as larger commercial farmers and many local elites may be primarily concerned with profits or social status.

Divergent perceptions

Vulnerable pastoralists' and peasants' perceptions of the most urgent and immediate threats to their livelihoods include alienation of traditional lands, insecure land rights, armed conflicts and brigandage, excessive rents, fees or taxes, exploitive markets and credits, absence of rudimentary public services and the like. Drought is usually perceived as a serious problem, but not one that can be averted by local efforts. Programmes designed to improve natural resource management have to deal with these diverse local realities and perceptions in dryland regions. UNEP's question about what dryland managers need in the way of assistance is not answerable in general terms.

The same is true concerning UNEP's question about the social impacts of drought and land degradation. Subordinate social groups with few resources will inevitably be among those most negatively affected. They may frequently blame drought and land degradation for their plight, but only as contributing factors. Deeper probing always brings out many other fundamental grievances such as those mentioned above.

Grassroots responses

Vulnerable pastoralists and cultivators in dryland regions are not passive fatalists when their livelihoods are threatened by drought and other adversities. They respond in many ways, but responses are shaped by their perceptions of the threats and opportunities confronting them. These may be rather different from those of government officials or of those in development and aid organizations.

Coping with adversity

The most common peasant response is to adapt customary production and consumption patterns to increasingly difficult circumstances. These so-called coping strategies frequently imply reduced food consumption leading to malnutrition and many other sacrifices in order to maintain enough productive assets to resume traditional production in less adverse times. They also often imply accelerated land degradation associated with over-use of inadequate natural resources. These abuses of land and health seldom arise from the peasants' carelessness or ignorance, but rather from necessity in order to survive.

Emergency relief programmes can sometimes alleviate immediate suffering. They will contribute little or nothing to better land management and longer-term protection of livelihoods, however, unless the root causes of vulnerability to drought and other adversities are attenuated. Programmes aimed at diminishing starvation and malnutrition are important for humanitarian reasons. They are only palliatives unless they also contribute to real sustainable development. This requires much more than locally focused emergency relief and land conservation efforts.

New livelihood sources locally

Another coping strategy has been for members of dryland vulnerable groups to find alternative sources of livelihood in or near their traditional communities. This often takes the form of engaging in petty trade or providing services to neighbours with more animals or crops. The total basic food and fuel supplies in the community are not increased as a result, but they may become more widely distributed.

Intensifying farming systems to produce more food for consumption or sale has become extremely difficult in land-degraded regions without considerable outside support. More intensive use of local natural resources often accelerates land degradation. Increased production and sale of cash crops frequently implies the use of purchased inputs such as seed and fertilizers, as well as investments in infrastructure such as transport and irrigation. Even then, these farming systems may not be sustainable because of unfavourable prices and land degradation. New rural industries creating gainful employment also require markets and outside investments that are always in short supply.

Only occasionally has generation of alternative sources of income locally offered a solution for those whose livelihoods are threatened in dryland regions by drought and the host of other adverse factors. There are exceptions where intensive cash cropping systems, ecotourism or rural industries have apparently prospered for long periods and could be sustainable. Explanations of these apparent successes have to be sought principally in the broader society.

Migration

A third response by members of vulnerable groups has been to migrate either temporarily or permanently. This has been a common strategy for maintaining livelihoods. It often contributed to land degradation somewhere else and implied many hardships.

Whether migrants should be considered as ecological refugees, or economic or political ones, is frequently impossible to determine except in an arbitrary way. In some cases, drought and land degradation may have been merely catalysts determining the date of departures that would have occurred in any case at some other time for reasons such as land alienation, indebtedness, war, political oppression or the quest for better opportunities. In other cases, desertification and drought may have been the major causes of out-migration that would otherwise have been indefinitely deferred.

Collective initiatives

A fourth response by peasants and pastoralists has been collective resistance to land alienation and related processes threatening their livelihoods. Resistance was sometimes combined with collective efforts to improve land management and productivity. Such collective actions on the part of vulnerable groups are extremely risky for those undertaking them as they may threaten the vested interests of local elites, of the state or of other more powerful social actors.

Collective initiatives are frequently violently repressed. Where vulnerable peasant groups were able to find allies in other sectors of the national society and internationally, they have occasionally been successful in defending their lands, at least temporarily. Key factors leading to favourable outcomes seem to have been autonomous democratic organization of the vulnerable groups, a supportive state, or at least one somewhat respectful of legal processes and basic human rights, and a civil society in which they could find allies.

The self-organization of vulnerable groups with the objective of increasing their control over resources and institutions directly affecting them is usually essential in order for the powerful to take their problems seriously. When they are organized with some degree of autonomy, the state and other social actors are likely to view them as potential allies or opponents whose interests have to be taken into account. One policy implication of this conclusion is the need to promote respect for basic human and civil rights together with democratic institutions and procedures accessible to vulnerable pastoralists and small cultivators.

Programmes to control desertification

Policies and programmes aimed at halting or reversing soil degradation have mostly focused on its so-called direct causes. For example, pastoralists were often prohibited to burn dryland pastures and grazing was often forbidden in wooded areas. Shifting cultivation was prohibited or penalized in some places. Afforestation of eroded lands was undertaken, sometimes on massive scales. Shelter belts were frequently planted. Agroforestry, contour ploughing, terraces, bunds and other soil conservation practices have been vigorously promoted. Efforts were made to stabilize sand dunes by planting resistant grasses, shrubs and trees. Wasteful irrigation practices resulting in salinization were discouraged. Large areas were set aside as parks or other protected areas. Boreholes were drilled to provide water for livestock and efforts were made to limit livestock numbers while improving their quality. Large irrigation and resettlement projects were frequently undertaken. This list could be lengthened.

Disappointing results

Several of these land improvement programmes were relatively successful according to land conservation or social criteria for limited areas and for limited time periods. They were only rarely successful, however, by both social and ecological criteria. Overall, they apparently had little impact on dryland degradation in developing countries. 'Successful' small projects have seldom been replicable over large areas. They have usually been unsustainable after active international or national financial and technical support diminished, ceased, or was diverted to other objectives.

Reasons for failure

The reasons for only very limited success have been varied. The absence of vulnerable groups' participation in setting the objectives and in subsequent programme administration is prominent among them. Frequently corruption or inadequate resources were blamed. An inherent bias of many desertification control programmes and policies in favour of elite groups and against the most vulnerable ones went hand in hand with the lack of real popular participation. Contradictory policies and institutions were prominent obstacles.

Projects and programmes were seldom designed to meet vulnerable participants' livelihood concerns. Instead, they emphasized land conservation objectives and outsiders' perceptions of the peasants' problems. Government and international support tended to be short term and unpredictable. The more profound social causes of dryland degradation were usually treated superficially or simply neglected.

Key societal issues

In an increasingly global economy, issues of dryland degradation cannot be dealt with adequately at only local levels. The principal social processes and institutions driving desertification and inhibiting improved resource management locally are rooted in their national and international context. In order to confront localized land degradation problems effectively, public policies and market forces have to become much more supportive of sustainable development efforts. Land tenure systems and wider social relations have to be modified in the same direction. Farming systems, technologies and economic structures all have to become increasingly sustainable. Effects associated with demographic changes have to be approached in a more realistic manner than they have been in the past.

Popularly based strategies

Effective desertification control requires national and international strategies that focus primarily on promoting sustainable development. National strategies will have to be popularly based, both in the sense of placing a high priority on the needs and aspirations of vulnerable social groups and of the real participation of these groups in formulating the strategy's objectives and in its execution when it directly impinges on their lives. International institutions, policies and strategies will also have to become more democratic and responsive to the needs of poor people in poor countries. This is utopian, but anything less will fail. Technocratic strategies that are not popularly based will inevitably become socially unsustainable.

Policies and market forces

Policies and market forces have to be considered together. Policies designed to improve dryland resource management will necessarily rely on market forces to some extent in order to achieve their objectives. On the other hand, market forces are guided by the policy and institutional frameworks within which they must operate. Market failures and policy failures are two sides of the same coin. Both policies and market forces will have to be directed towards promoting poverty alleviation, social solidarity and ecologically sustainable development as primary goals.

It is impossible to generalize about what policies should be adopted in specific countries to improve natural resource management as each situation is different. National strategies should, among other things, attempt to provide real opportunities and incentives for vulnerable pastoralists and peasants as well as for better-off operators in dryland regions to manage their natural resources sustainably. At the same time, they have to improve livelihoods. Price, trade, credit, investment, social, fiscal and other policies all have to be designed to contribute towards meeting the priority objectives of sustainable development. The crucial point is that such strategies will have to be popularly based.

International policies also have to be modified. Pressures on poor countries and poor people to overexploit natural resources in order to increase exports are enormous. Writing off unproductive odious foreign debts could help to bring relief, so too could a meaningful code of conduct requiring transnational corporations to observe minimum ecological and social norms in their investments, production and trade. More serious attention could be given by international organizations to developing accounting systems both for governments and transnational enterprises that capture environmental and social costs associated with industrial production and consumption patterns, natural resource degradation, poverty and the like. Structural adjustment programmes could be given both a more human and a greener face. Additional international resources could be mobilized and channelled to support sustainable development strategies. Particularly important could be the evolution of more effective and democratic institutions capable of resolving international conflicts peacefully. It should then become possible to divert much of the some 5% of the world's total economic product now devoted to armaments to the social goals implied by sustainable development. This wish list too could be indefinitely extended.

Land reforms

Any sustainable development strategy would have to deal with issues of land tenure in dryland regions. In most countries, insecure and inequitable land tenure systems are among the principal factors contributing to dryland degradation. Land reforms aimed at providing equitable access to land by those actually working it and providing them with secure clear rights (and obligations) associated with their land tenures are essential, although not sufficient, for improving natural resource management. International organizations could be much more effective than they have been in working to help bring about such reforms.

The kinds of land reforms most appropriate for specific countries and situations have to be worked out locally in consultation with vulnerable agrarian groups and the other social actors. The principles of secure and equitable access with clear rights and obligations provide general guiding norms. Land tenure implies a sub-set of social relations. The national state has to play a crucial role in the existing 'world order' ('disorder'?) as it is the 'sovereign state' that ultimately sets the rules concerning property rights and arbitrates their application. Of course, this assumes a state with sufficient legitimacy and political resources to play this role, which is by no means always the case in practice.

Many dryland regions are located in countries where customary communal tenure systems still operate in practice. Nationally, however, the property regimes legally recognized by these states are primarily those of private and state ownership. This introduces intolerable uncertainty of tenure for participants of communal property regimes as their customary rights may be rather arbitrarily abrogated at any time. This is presently the case in much of Sub-Saharan Africa as well as a few regions in Asia and the Americas. Moreover, customary communal systems frequently have functioned rather well in providing equity of access to land as well as clear rights and obligations for their members. They were particularly well suited for managing low-value range and woodlands where costs of establishing cadastrals, land registries and other requisites for effective private property or state property regimes would be prohibitive as well as being very questionable for social reasons.

A priority reform in regions where customary communal tenure systems are still vigorous would be for the state to recognize customary land rights. This would provide greater security for community members from the danger of encroachment of their land rights by private interests or by the state. Of course, this would have to be accompanied by norms to guarantee continued democratic participation by the communities' members and to prevent excessive concentration in the control of communal lands by small elites. These rules would preferably be worked out with the communities concerned.

In dryland areas where customary tenure systems have already broken down or never existed, other kinds of land reform measures have to be envisioned. Clear and equitable rights and obligations are possible under communal, private and state property regimes.

Sustainable farming systems

Policy reforms and land tenure reforms could greatly facilitate the emergence of more sustainable farming systems in dryland regions. The literature suggests that, to the extent possible, such farming systems should be based on indigenous knowledge and the use of local resources in each locality. Growing integration into national and international markets, however, inevitably implies selective use of purchased inputs and technologies.

There should be greater support for site-specific research aimed at integrating traditional indigenous knowledge and practices with the insights of modern science into ecologically and socially sustainable farming systems. Such systems should be developed with the participation of the local groups who will use them. Pastoralists and small-scale farmers will usually want to find production systems that improve their livelihoods without excluding members of their communities. Sustainable farming systems that use available labour productively while economizing in the use of externally purchased inputs would be the ideal. Constraints of resources, markets and price relations often make attainment of this ideal difficult or impossible. Creation of off-farm productive employment would remain a major issue in most dryland regions.

Eventually there will have to be major changes in production and consumption patterns and technologies throughout both the industrial and industrializing world. Ever-growing demands for natural resources are generated by industrial production and consumption systems in the developed countries and increasingly in developing ones. These put additional pressures on natural resources in dryland regions. These pressures are transmitted from industrialized countries and urban centres in poor countries to dryland areas through institutions, policies and market forces. Additional environmental stress results from industrial pollution and greenhouse gases. These are global issues that require international attention.

Demographic trends

Population growth also contributes to pressures on natural resources in dryland regions. The stress on land and water resources attributable to demands generated by the poor, however, is very small in comparison with those attributable to industrial activities and to the consumption patterns of the rich. Consumption of natural resources by the very poor is rather insignificant in most countries. Within dryland regions, the relationships between population dynamics, natural resource degradation and development are extremely complex and often contradictory. It is simplistically misleading to treat population growth and poverty as the principal root causes of desertification, as is often done.

Population projections for the next three or four decades at national and global levels have to be taken very seriously. Migrations make them much less reliable locally. Most demographic growth will take place in developing countries.

Present rates of growth imply a doubling of the world's population during the next four decades. This rate of demographic growth would be unsustainable if it were to continue during the twenty-first century, or if it were to decline only slightly. On the one hand, if the dominant development style continues to leave a majority in poverty while only a minority become better off, this would inevitably lead to social chaos. On the other hand, if the vast majority of the world's population were to adopt present industrial country production systems and life styles, the pressures on the natural environment would become intolerable.

There are clearly limits to demographic and economic growth as they have been conventionally defined. The only way out of this dilemma is for the content and meaning of development to change in practice in both developed and developing countries. In the future, quality will have to take precedence over quantity.

An important implication of this conclusion is that real social development is an indispensable component of any sustainable development strategy. There is no other acceptable way to bring down population growth rates. Increasing security, living levels, education and opportunities of the poor is essential for achieving a stable world population. Fortunately, the same type of social development that could eventually check demographic growth could also make sustainable management of natural resources in dryland regions a much more feasible goal.

NOTES

1. Some readers have questioned this statement. Measures of land productivity imply a ratio of socially desired output (product) to an area of land (input). Outputs can be expressed in physical terms such as tons or calories of food and biomass, numbers of diverse species and the like. Alternatively, productivity can be estimated in qualitative terms such as aesthetic or moral and philosophic satisfactions, or in economic terms of gross or net monetary values per hectare. In today's global economy there is a growing tendency to estimate land productivity in monetary values that are linked to commodity prices in international markets. This does not make economic valuations any more scientific or less socially determined than are alternative measures. In any event, markets are constantly changing. Even when there is some agreement about what measures of productivity to use, however, the technical problems remain extremely controversial. There are several good reviews of some of these technical difficulties in estimating land degradation (Blaikie, 1985; Horowitz, 1990).
2. UNCTAD cites an OECD study estimating that in OECD countries transfers from tax payers and consumers (through higher prices) to support their agricultural policies amounted to $299 billion in 1990. These policies undoubtedly helped to improve the livelihoods of their remaining farmers, although very unequally. They also generated huge market surpluses that were very expensive to manage. Using the same perverse logic that led them to conclude that most developing countries are underpolluted, narrowly focused neo-liberal economists might even argue that in a world of unmarketable agricultural surpluses and high farm subsidies there is no net current economic cost to society from lower agricultural production on degraded drylands.
3. There are several other good case studies. An outstanding one is Watts' (1987) research with the Hausa in northern Nigeria.
4. (Another underestimated and underdeveloped area is the commercialization of wild dryland plant products (Stiles, 1988)—Ed.)

REFERENCES

Adams, P. (1991) *Odious Debts*. London, Earthscan.

Allen, T. (1993) Social and Economic Aspects of Mass 'Voluntary' Return of Refugees from Sudan to Uganda between 1986 and 1992. Draft report for UNRISD, February.

AVANCSO (Asociaciòn para el Avance de las Ciencias Sociales en Guatemala). (1992) *Donde està el Futuro? Procesos de Reintegraciòn en Comunidades de Retornados*. Guatemala City.

Barraclough, S.L. (1991a) *An End to Hunger? The Social Origins of Food Strategies*. London, Zed Books.

Barraclough, S.L. (1991b) Migration and development in rural Latin America. In: Mollet, J.A. (ed.), *Migration in Agricultural Development*. London, Macmillan Academic and Professional Ltd.

Barraclough, S.L. and K. Ghimire (forthcoming) *Forests and Livelihoods: The Social Dynamics of Deforestation in Developing Countries*. London, The Macmillan Press Ltd.

Barrow, C.J. (1991) *Land Degradation: Development and Breakdown of Terrestrial Environments*. Cambridge, Cambridge University Press.

Beckerman, W. (1974) *In Defence of Economic Growth*. London, Cape.

Beckerman, W. (1992) Economic growth and the environment: 'Whose growth? Whose environment?' *World Development* **20**(4) April.

Bedrani, S. (1983) Going slow with pastoral cooperatives: Reversing the degradation of the Algerian steppe is an awkward, arduous task. *CERES* **16**(4) July–August.

Behnke, R.H. and I. Scoones (1992) *Rethinking Range Ecology: Implications for Rangeland Management in Africa*. World Bank, Environment Working Paper No. 53, Washington, DC.

Beinart, W. (1984) Soil erosion, conservationism and ideas about development: A Southern African exploration, 1900–1960. *Journal of Southern African Studies* **11**(1) October.

Blaikie, P. (1985) *The Political Economy of Soil Erosion in Developing Countries*. London, Longman.

Blaikie, P. and H. Brookfield (1987) *Land Degradation and Society*. London, Methuen.

Bonfiglioli, A.M. (1992) *Pastoralists at a Crossroads: Survival and Development Issues in African Pastoralism*. NOPA, UNICEF/UNSO Project for Nomadic Pastoralists in Africa.

Brown, L. (1973) *Population and Affluence: Growing Pressure on World Food Resources*. Paper 15, Overseas Development Council, Washington.

Brown, L. (1979) Where has all the soil gone? *Mazingira* **10**: 61–68.

Buchanan-Smith, M. (1990) *Drought and the Rural Economy in Botswana: An Evaluation of the Drought Programme, 1982–1990, Drought, Income Transfers and the Rural Household Economy*. Food Studies Group, Queen Elisabeth House, University of Oxford.

Buchanan-Smith, M. and M. Kelly (1991) *North Sudan in 1991: Food Crisis and the International Relief Response*. Draft, Institute of Development Studies, Brighton.

Buchanan-Smith, M. and C. Petty (1992) *Famine Early Warning Systems and Response: The Missing Link?, Sudan: 1990–1991*. Draft prepared for Save the Children Fund, London and Institute of Development Studies, Brighton.

Cavalcanti, C. (1989) Dimension socio-économique de la sécheresse de 1979–80 dans le Nordeste du Brésil. In: Bret, B. (ed.), *L'Homme Face aux Sécheresses*. Paris, IHEAL.

CEPAL (Comision Económica para América Latina y el Caribe) (1992) *Balance Preliminar de la Economía Latino américana*. Santiago de Chile.

CIDA (Comité Interamericano para el Desarrollo Agrícola) (1966a) *Tenencia de la Tierra y Desarrollo Socio-Económico del Sector Agricola: Peru*. Secretaría General de la Organización de los Estados Americanos, Washington, DC.

CIDA (1966b) *Land Tenure Conditions and Socio-Economic Development of the Agricultural Sector: Brazil*. Pan American Union, Washington, DC.

Cliffe, L. (1988) The conservation issue in Zimbabwe. *Review of African Political Economy* No. 42.

Dahl, G. and A. Hjort (1979) *Having herds: Pastoral herd growth and household economy*. Stockholm Studies in Social Anthropology, No. 2, University of Stockholm, Stockholm.

Daly, H. and R. Goodland (1992) *An Ecological Assessment of Deregulation of International Commerce under GATT*. Discussion Draft Environment Working Paper, The World Bank, Washington, DC.

Dasgupta, P. (1982) *The Control of Resources*. Oxford, Basil Blackwell.

Davies, S., M. Buchanan-Smith and R. Lambert (1991) *Early Warning in the Sahel and Horn of Africa: The State of the Art, a Review of the Literature*, Vol. 1. Institute of Development Studies, University of Sussex, Brighton.

Diegues, A.C. (1992) *The Social Dynamics of Deforestation in the Brasilian Amazon: An Overview*. UNRISD Discussion Paper No. 36, Geneva.

Dregne, H.E. (1976) *Soils of the Arid Regions*. Amsterdam, Elsevier.

Droulers, M. (1989) Les réponses paysannes au problème de la sécheresse dans le Sertâo Nordestin. In: Bret, B. (ed.), *Les Hommes Face aux Sécheresses*. Paris, IHEAL.

Erlich, P.R. (1992) The value of biodiversity. *Ambio* **XXI** (3).

Erlich, P.R. and A.H. Erlich (1970) *Population, Resources, Environment: Issues in Human Ecology*. San Francisco, CA, W.H. Freeman.

Frankie, R. and B.H. Chasin (1980) *Seeds of Famine*. New York, Universe Books.

Fuller, T.D. (1987) Resettlement as a desertification control measure: A case study in Darfur region, Sudan. *Agricultural Administration and Extension* **25**(4).

Gadgil, M. and R. Guha (1992) *This Fissured Land: An Ecological History of India*. Delhi, Oxford University Press.

Garcia, R. (1984) *Nature Pleads Not Guilty, Drought and Man*, Vol. 1, IFIAS, Oxford, Pergamon Press.

Garcia, R. and J. Escudero (1982) *The Constant Catastrophe*, Vol. 2, Oxford, Pergamon Press.

Garcia, R. and P. Spitz (1986) *The Roots of Catastrophe*, Vol. 3, Oxford, Pergamon Press.

Garcia Barrios, R. and L. Garcia Barrios (1992) Environmental and technological degradation in agriculture: A consequence of rural development in Mexico. *World Development* **XVIII** (II).

Ghimire, K. (1993) *Population Dynamics, Environmental Changes and Development Processes: Lessons from Costa Rica, Pakistan and Uganda*. Geneva, UNRISD.

Glantz, M.H. (1984) Floods, fires and famine: Is El Niño to blame? *Oceanus* **27**(2).

Glantz, M.H. (ed.) (1987) *Drought and Hunger in Africa: Denying Famine a Future*. Cambridge, Cambridge University Press.

Glantz, M.H. (1988) Drought follows the plow. *The World & I* April.

Glantz, M.H. (1989) Drought, famine and the seasons in Sub-Saharan Africa. In: Huss-Ashmore, R. and S. Katz (eds), *African Food Systems in Crisis, Part One: Microperspectives*. New York, Gordon and Breach.

Goodland, R. and H. Daly (1993) *Poverty Alleviation is Essential for Environmental Sustainability*. The World Bank Environmental Division Working Paper, 1993–42, Washington, DC.

Grainger, A. (1988) Estimating areas of degrading tropical lands requiring replenishment of forest cover. *International Tree Crops Journal* No. 5.

Grainger, A. (1990) *The Threatening Desert*. London, Earthscan.

Gray, L. and M. Kevane (1993) For whom is the rural economy resilient? Initial effects of drought in Western Sudan. *Development and Change* **24**.

Hall, A.L. (1983) *A Re-Appraisal of Government Irrigation in North-East Brazil*. Mimeo, London School of Economics, London.

Hammer, T. (1993) *Environmental Projects—Power and Policy: Structural Factors Influencing People's Participation in Natural Resource Management, Some Cases from Sudano-Sahelian Africa*. Draft, Geneva, UNRISD.

Hjort af Ornas, A. and M.A. Mohamed Salih (eds) (1989) *Ecology and Politics: Environmental Stress and Security in Africa*. Scandinavian Institute of African Studies, Uppsala.

Horowitz, M. (1990) Donors and deserts: The political ecology of destructive development in the Sahel. *African Environment* Nos 25–28, **VII**, 1–4, Dakar.

Hutchinson, R.A. (1991) *Fighting for Survival: Insecurity, People and the Environment in the Horn of Africa*. IUCN Sahel Programme Study, IUCN, Gland.

Ibrahim, F.N. (1982) The role of women peasants in the process of desertification in Western Sudan. *GeoJournal* **6**(1).

Ibrahim, F.N. and H. Ruppert (1991) The role of rural–rural migration as a survival strategy in the Sahelian zone of Sudan. *Geojournal* **24**(1).

IFAD (International Fund for Agricultural Development) (1986) *Soil and Water Conservation in Sub-Saharan Africa: Issues and Options*. A report prepared for the International Fund for Agricultural Development by the Centre for Development Cooperation Services, Free University, Amsterdam.

IFAD (1992) *Soil and Water Conservation in Sub-Saharan Africa: Towards Sustainable Production by the Rural Poor*. A report prepared for the International Fund for Agricultural Development by the Centre for Development Cooperation Services, Free University, Amsterdam.

Independent Commission on International and Humanitarian Issues (1985) *Famine: A Man Made Disaster?* London, Pan Books.

Jodha, N.S. (1982) The role of administration in desertification: Land tenure as a factor in the historical ecology of Western Rajasthan. In: Spooner, B. and H.S. Mann (eds), *Desertification and Development: Dryland Ecology in Social Perspective*. London, Academic Press.

Johnson, A. (1973) *Sharecroppers of the Sertâo*. Stanford, Stanford University Press.

Johnson D.H. and D.M. Anderson (eds) (1988) *Ecology of Survival: Case Studies from Northeast African History*. Boulder, CO, Westview Press.

Johnson, D.H. and W. Ofosu Amaah (1974) Changing patterns of pastoral nomadism in North Africa. In: Hoyle, B.S. (ed.), *Spatial Aspects of Development*. Chichester, John Wiley.

Keen, D. (1992) A Political Economy of Refugee Flows from South-West Sudan, 1986–1988. *UNRISD Discussion Paper No. 39*, UNRISD, Geneva, November.

Lane, C. (1990) Barabaig Natural Resource Management: Sustainable Land Use under Threat of Destruction. *UNRISD Discussion Paper No. 12*, UNRISD, Geneva, June.

Lane, C. (1993) The Barabaig/NAFCO conflict in Tanzania: On whose terms can it be resolved? *Forests, Trees and People Newsletter* No. 20, April.

Lashova, G.A. (1989) The crisis of nomadic farming and problems of desertification in North African countries. *Problems of Desert Development* **3**.

Little, P.D. (1985) Absentee herd owners and part-time pastoralists: The political economy of resource use in Northern Kenya. *Human Ecology* **13**.

Livingstone, I. and M. Assunçâo (1989) Government policies towards drought and development in the Brazilian Sertâo. *Development and Change* **20**(3).

Markakis, J. (1993) *Conflict and the Decline of Pastoralism in the Horn of Africa*. Institute of Social Studies. London, Macmillan Press.

Mascarenhas, A. (1993) *Ecology and Deforestation in Semi-Arid Kondoa District—Tanzania*. Draft prepared for UNRISD, July.

Meadows, D.H., D.L. Meadows and J. Anders (1970) *The Limits to Growth*. London, Earth Island.

Monod, T. (ed.) (1975) *Pastoralism in Tropical Africa*. Oxford, Oxford University Press.

Nelson, R. (1988) *Dryland Management: The 'Desertification' Problem*. Environment Department Working Paper 8, The World Bank, Washington, DC.

Nelson, R. (1990) *The Management of Drylands*. Unpublished manuscript prepared for the World Institute for Development Economics Research (WIDER), Helsinki.

Reed, David (ed.) (1992) *Structural Adjustment and the Environment*. London, Earthscan.

Rhodes, S.L. (1991) Rethinking desertification: What do we know and what have we learned? *World Development* **19**(9).

Salih, M.A.M. (1985) Pastoralists in town: Some recent trends in pastoralism in the northwest of Omdurman District (Sudan). *Pastoral Development Network Papers* No. 20b, ODI, London.

Savané, M.-A. (1986) *Femmes et Développement en Afrique de l'Ouest*. Geneva, UNRISD.

Savané, M.-A. (1992) *Populations et Gouvernements Face aux Problèmes Alimentaires: Regards sur des Zones de l'Afrique de l'Ouest*. Geneva, UNRISD.

Schneider, S.H. (1989) The greenhouse effect: Science and policy. *Science* **243**, February.

Simon, J.L. (1981) *The Ultimate Resource*. Oxford, Martin Robertson.

Somerville, C.M. (1986) *Drought and Aid in the Sahel, a Decade of Development Cooperation*. Boulder, CO, Westview Press.

Sousans, J., B. Shrestha and L.P. Uprety (1993) *Trees, People and Governments: The Social Dynamics of Deforestation and Sustainable Forest Management in Nepal*. Draft, Geneva, UNRISD.

South Commission (1990) *Challenge to the South*. Oxford.

Speirs, M. and O. Olsen (1992) *Indigenous Integrated Farming Systems in the Sahel*. World Bank Technical Paper No. 179, Washington, DC.

Spooner, B. and H.S. Mann (1982) *Desertification and Development: Dryland Ecology in Social Perspective*. London and New York, Academic Press.

Stiles, D. (1983) Desertification and pastoral development in northern Kenya. *Nomadic Peoples* **13**.

Stiles, D. (1987) Classical versus grassroots development. *Cultural Survival Quarterly* **11**(1).

Stiles, D. (1988) Arid land plants for economic development and desertification control. *Desertification Control Bulletin* **17**.

Stiles, D. (1992) The Gabbra: Traditional social factors in aspects of land-use management. *Nomadic Peoples* **30**.

Swift, J. (1973) Disaster and a Sahelian nomad economy. In: Dalby, K. and R.J. Harrison (eds), *Drought in Africa*. Centre for African Studies, University of London, London.

Toulmin, C. (1993) Combating Desertification: Setting the Agenda for a Global Convention. *IIED Issue Paper* No. 42.

Treacy, J.M. (1989) Agricultural terraces in Peru's Colca valley: Promises and problems of an ancient technology. In: Browder, J.O. (ed.), *Fragile Lands of Latin America, Strategies for Sustainable Development*. Boulder, CO, Westview Press.

UNCED (United Nations Conference on Environment and Development) (1992) *Agenda 21*. Geneva.

UNCTAD (United Nations Conference on Trade and Development) (1991) *Trade and Development Report*. New York, United Nations.

UNEP (United Nations Environment Programme) (1991) *Status of Desertification and Implementation of the United Nations Plan of Action to Combat Desertification*. UNEP, Nairobi, UNEP/GCSS III/3.

UNEP (1992) *World Atlas of Desertification*. London, UNEP and Edward Arnold.

Utting, P. (1993) *Trees, People and Power*. London, UNRISD/Earthscan.
Watts, M. (1987) Drought, environment and food security: Some reflections on peasants, pastoralists and commoditization in dryland West Africa. In: Glantz, M.H. (ed.), *Drought and Hunger in Africa: Denying Famine a Future*. Cambridge, Cambridge University Press.
Weiss, H. *et al.* (1993) The genesis and collapse of third millennium north Mesopotamian civilization. *Science* **261**, 20 August.
Wilson, E. (ed.) (1988) *Biodiversity*. Washington, DC, National Academy Press.
World Bank (1991) *The Forest Sector*. A World Bank Policy Paper, Washington, DC.
World Bank (1992) *World Development Report 1992*. New York, Oxford University Press.

Part II

PARTICIPATORY APPROACHES AND METHODS

'Community participation', 'participatory research', 'participatory rural appraisal', 'the bottom-up' approach and so on are terms being increasingly used by NGOs, donor agencies and even governments. They feature prominently in Chapter 12 of Agenda 21, which deals with desertification, and in the Desertification Convention. But 'participation' is not a panacea for the problems, as the following chapters warn.

People do have to solve their own problems, however. Top-down approaches have not worked. Various methods have been developed and are being tried out to facilitate people being able to conceptualize their problems and plan for remedial action to improve natural resource management. This part presents various approaches and methods that have been and are being tried, and highlights some of the successes and problems gained from experience.

3 The Active Method of Participatory Research and Planning (MARP) as a Natural Resource Management Tool

MAMADOU BARA GUÈYE
Ecole Nationale d'Economie Appliquée, Dakar, Senegal

INTRODUCTION

The promotion of community participation in natural resource management is currently a central concern for natural resource management projects. This is primarily due to the fact that the strategies implemented thus far have not been very effective, because of their essentially technological bias. The alarming scale of degradation of the economic production base in rural areas in Africa has rendered the implementation of long-term, viable natural resource management mechanisms all the more urgent. In addition, the resulting scarcity of exploitable resources (in particular, land) has made it harder to gain access to, and to manage, these resources. Attempts at regulation and management by government bodies have resulted in the destruction of traditional management mechanisms, thereby creating conflict situations which are often difficult to resolve.

Awareness of the present fragility of resources is very high within the village communities, which are endeavouring to implement more appropriate management models. Regrettably, these initiatives are not always taken into account and made use of in the definition of an institutional framework for the management of natural resources. It now seems to be increasingly accepted that the institutional natural resource management projects and programmes cannot hope to safeguard local resources in the long term because of their own limited lifetime. Such guarantees can only be offered by permanent community structures, fully empowered to do so. But the big question remains: what can be done to promote the genuine participation of populations in natural resource management?

To take the specific example of the Sahel, the village land management approach has been adopted by several natural resource management projects,

Social Aspects of Sustainable Dryland Management. Edited by Daniel Stiles

because of the scale on which the village land management approach is applied. Village lands represent a more relevant and more manageable space to local people than the scale on which most major development projects are usually implemented. Consequently, the village land management approach has the potential to promote much wider participation by local people. The active method of participatory research and planning (Méthode Active de Recherche et de Planification Participative: MARP) currently offers a range of tools and techniques facilitating the village land management approach.

THE ACTIVE METHOD OF PARTICIPATORY RESEARCH AND PLANNING (MARP)

Definition of MARP

MARP is a method of participatory research and planning using visual tools developed by the populations themselves. The use of these tools promotes development of the know-how of local populations and facilitates communication between the village community and technicians. The method is iterative, multidisciplinary and flexible. Where natural resource management is concerned, MARP tools are increasingly being used by local people in planning and managing relevant programmes. MARP thus represents a method which aims to give such people the necessary tools for the implementation of community programmes. As such, its scope extends further than that of mere research. The tools used include:

Resource maps
Sociological maps
Matrix classifications
Venn diagrams
Flow charts
System charts
Seasonal calendars
Historical analyses
Analyses of the village land or transect
Classifications by level of wealth

The following are important principles of MARP.

Making use of local knowledge

One of the important methodological principles of MARP is the use of the knowledge of local populations. The process draws on the actual experience of the populations themselves. Where natural resource management is concerned, increased efforts by projects to take into consideration the knowledge gathered by local populations could offer an important short-cut in the elaboration and implementation of more appropriate natural resource management programmes. In addition, experience has shown that use of MARP in the participatory planning of natural resources can lead to a much greater harnessing of local knowledge.

Three-pronged approach

The paticipatory planning process includes research. An essential issue in any research process, participatory or otherwise, is the reliability of the instruments and, by extension, the validity of the results achieved at the end of the process. One constant concern in the application of MARP is to ensure that the short time spent in the field does not constitute a factor affecting the quality of the results. It is therefore essential that work-time should be rationally managed, and the principle of 'optimal ignorance' dictates that the work-time should be devoted to acquiring knowledge of direct relevance to the objectives of the study. Only too often research projects approach local people to obtain information that will be only partially used, if at all.

The three prongs of the approach are:

(1) Multidisciplinarity, which makes it possible to analyse the phenomena being studied from different points of view, giving a better reflection of the very complex realities of rural life.
(2) Diversification of tools and techniques, so that different units of measure are used for each phenomenon being studied. This form of blending definitely enhances the quality of the information obtained.
(3) Diversification of information sources, in order to avoid a sectoral view of reality. In addition, certain biases inherent in different social groups can either be eliminated or at least reduced.

Role reversal

MARP is based on a new paradigm of development, one of the bases of which is that technicians must change their attitude. This change in attitude must be manifested through role reversal. This means that, in the MARP process, local people are now analysts and presenters of their own situations while the technicians play the role of catalysts and facilitators. This process of change can sometimes be rather difficult since it requires technicians to surrender some of their powers to local people. Hence it can never be stated too often that MARP is designed not for the technicians but for the local people themselves. For this reason, training of the technicians should be considered as nothing more than an interim stage in the transfer of tools and techniques to grassroots communities.

Iteration

MARP is an iterative learning process and not merely a data-collection exercise. This makes the process rather complex but, at the same time, facilitates utilization of the skills of those participating in the process.

Strengths of MARP

MARP helps local people to analyse their own situations. Experiments in the use of MARP in the field demonstrate the great facility with which people use visual tools and techniques as aids in analysing situations of direct relevance to themselves.

Visualization facilitates the transfer to local people of research findings. In most village communities, illiteracy impedes the use of written information. With MARP it is much easier for a population to apply the results, since the process relies on visual aids developed by the people themselves and relating to local situations.

Visualization considerably reduces the marginalization of certain groups in the analysis process. A number of examples gathered across Africa demonstrate the great strength of MARP tools as aids facilitating the participation by certain traditionally marginalized groups in the planning and implementation of various actions. While generalizations should be avoided, it has nonetheless been noted that the process itself was so exciting that, by the end, it had engaged the attention of all groups. One thing should be stressed, however: the importance of ensuring that the participation of these groups in joint discussion exercises is matched with their participation in the decision-making process.

Visualization facilitates communication between local people and technicians. It is now generally accepted by their users that MARP tools are equally effective aids for communication and extension work. The potential of MARP as an aid for rural extension work has still not been fully explored.

POTENTIAL DANGERS OF MARP

Too popular, too soon!

The rapid spread of MARP also represents a fairly sizeable risk. The demand for support generally exceeds the response capacity of the qualified staff available. The usual consequence of this is that institutions apply the approach without making the necessary methodological preparations. In fact, the apparent flexibility of MARP also makes it a very demanding method at both the intellectual and physical levels. MARP produces bad results when it is badly used. This problem currently represents one of the most serious threats for the development of MARP.

Expectations aroused

MARP arouses high expectations with local populations. The practical process is so exciting and so catches the attention of local populations that they invariably—and with good reason—expect that its results will provide the basis for future field action. For this reason, MARP must always be linked to ongoing programmes or actions in the field. Unfortunately, their failure properly to grasp the end-purpose of this method leads certain users to give pride of place to the research itself. The frustration caused thereby can undermine the interest which populations might otherwise take in the method.

Extractive use

MARP is, first and foremost, a planning tool designed for the populations themselves. But there is currently a pronounced tendency to turn MARP into a mere research method. This creates the danger that the method will become distorted since the participation of populations will be of more benefit to some outside institution or individual.

MARP as cure-all

Inadequate understanding of MARP is likely to turn it into an alternative method which excludes all others. MARP works in certain contexts and not in others. Some organizations see MARP as a miracle method which offers a remedy for any difficulty encountered in the field. A method is only as good as the use to which it is put, however; tools are nothing more or less than the means to an end. Consequently, the method should not be isolated from other methods and approaches, whose possible complementarity with MARP should be wisely exploited.

Lack of monitoring

The rapid development of this method makes it difficult to devise any mechanism to monitor the use of its tools. If the potentially disastrous consequences of the cumulative effects of incorrect use are to be avoided, the various experiments must be regularly and thoroughly monitored. Unfortunately, this requirement does not seem to have been properly appreciated by all MARP users. One major constraint in dealing with this problem is that no institution can claim the right to exercise this function, still less impose any standard of reference. This undoubtedly increases the risk of the method being used incorrectly.

VILLAGE LAND MANAGEMENT

In the specific context of the Sahel, the main instrument for natural resource management at the local level is the village land management approach (*approche gestion de terroirs villageois*). This approach earned its credentials in the late 1970s. It can be defined as the total set of institutional and organizational mechanisms implemented by a given community or group of communities with a view to the sound and sustainable management of the natural resources within an area which they consider to be theirs.

The methodological and practical attraction of this approach consists in its departure from the official systems for the management of village land based on spatial demarcations which do not necessarily correspond to areas relevant to local communities. The village land represents a territory which local communities can control more easily, particularly in the implementation of natural resource management programmes. It could be argued, therefore, that the village land management approach is potentially conducive to increased participation by local people provided the systems set in place recognize their right to control local resources.

The village land management approach combines the planning and the management of local resources. Its implementation comprises the following basic stages:

- Discussion and diagnosis with a view to identifying problems related to the management of resources
- Establishment of a local organization for the management of local resources
- Demarcation of the village land
- Elaboration of a management plan, based on zoning and co-ordinated with the local organization entrusted with management of the resources
- Implementation of the plan
- Monitoring and evaluation of the plan
- Report.

WHY LINK VILLAGE LAND MANAGEMENT AND MARP?

Although the implementation stages of this approach have been clearly identified, it is still hampered by methodological constraints caused by the lack of specific tools and techniques for its application. This largely explains why different institutions currently involved in village land management use different methods for its application. Methodological analysis and practical experiments involving the possible linking of village land management and MARP are among the efforts made to fill this void. Experience gathered elsewhere (in particular, Asia) demonstrates the usefulness of MARP tools as aids in the implementation of community planning and management programmes for natural resources. In India, for example, patient and carefully planned efforts by certain local non-governmental organizations have enabled some village communities to make independent use of the MARP process in defining and implementing programmes for the management of their village lands.

At the same time, the village land management approach, as currently implemented, still has certain limits:

(1) The village land management approach takes a restrictive view of village land, which it considers as a finite spatial entity. In fact, it is well known that, in reality, the boundaries of African agro-systems are very fluid. Several village lands can overlap, resulting in highly complex management mechanisms. In the Sahel in particular, it has been noted that desertification has considerably changed traditional natural resource management practices. Neighbouring communities are increasingly pooling their natural resources and developing intercommunal management mechanisms which are better adapted to the new environmental conditions. When MARP is applied to village land management, the use of its tools and techniques should be adapted to these realities, which, fortunately, can be analysed using various MARP techniques.
(2) Village land management programmes take insufficient account of the needs of certain groups, whose activities are not confined within a given area. Such is the case, in particular, with pastoralists. This often leads to the exacerbation of conflicts involving access to certain key resources. Efforts must be made to refine the methodology so that MARP can be more effectively applied to the realities of pastoral life. In certain cases, new ground must be broken to find new tools.
(3) Current legislation on the management of natural resources is often in conflict with the spirit of the village land management approach. The participation required from local populations is only possible and justified to the extent that these populations have a certain measure of control over local resources and are given assurances that they will have equal control over the use to which the results of their efforts are put.
(4) Certain projects seem not to be interested in linking implementation at the village land level to an overall development plan. The village land is not an isolated entity, however, and both economic policies and the immediate socio-economic environment influence local processes. Consequently, it is important to 'think globally and act locally'.
(5) From looking at village land management projects, it is clear that in many of them participation by the local people is more a means than an end. The aim is usually to generate community participation in order to ensure the success of the activities under way, even though the overall programme being implemented may not give any indication of how skills and powers are actually going to be transferred to the community. When participation is only a means to an end, its long-term viability will depend on the continuance of the activities that brought it about in the first place. There is a risk that participation will end with these activities.

POSSIBLE APPLICATIONS OF CERTAIN MARP TOOLS IN NATURAL RESOURCE MANAGEMENT

Resource maps

The resource map is an effective tool for making inventories of resources. Maps drawn up by local populations may be very detailed and go beyond inventorying available resources, providing a basis for the zoning of village land. This tool can also lead to a better understanding of the modalities by which land is appropriated and the rules governing its circulation among members of the community. The resource map therefore represents a useful tool both for planning and for management.

Analysis of the village land or transect

This supplements the resource map and adds participatory observation to the process of visualization. With the use of this tool, the multidisciplinary team (composed either of technicians and villagers or exclusively of villagers) can better appreciate visualized elements in the resource map drawn up by the villagers.

Sociological map

This is a useful and effective tool of inventorization and classification. In resource management it can be used to classify the different activities or groups in accordance with their level of access to land resources; it also gives information about the availability and state of the resources. In this way, it constitutes an important decision-making tool in the area of land management.

Venn diagram

The Venn diagram gives information about the traditional or modern institutions involved in the management of local resources. It also sheds light on the impact of these different structures as perceived by the local people and the ways they are interrelated. In the planning of natural resource management activities, the Venn diagram is useful for analysing the roles to be played by the different institutions.

Flow charts

The range of MARP tools has been expanded with the addition of flow charts describing land-tenure relationships. These usefully supplement the Venn diagram by giving more detailed information about the relationships between a given community and its environment (the lending or borrowing of land, land-tenure conflicts, co-management of resources, etc.). This tool is of vital importance, since some institutional models for the management of village land are vitiated by an assumption—usually mistaken—that village land has fixed limits.

System charts

System charts demonstrate the great intricacy of land-tenure systems in Africa. Thanks to the constant methodological refinement of these charts, the operation of land-tenure systems is now much better understood. They are highly analytical tools.

Seasonal analyses

Seasonal analyses show the fluctuations that occur within a given year or over the course of several years. When applied to the use of labour over the course of a year, they help identify the extent of pressure on the labour force at certain times of the year, so that village land management activities can be better programmed.

An entire range of tools for analysis and historical trends enable local populations to look back at previous practices and see to what extent these practices have been responsible for their present situation. Such trend analyses enable people to make forecasts about the 'possible future'. The importance of such a tool in programmes for the planning of natural resources is self-evident.

PRECONDITIONS

The long-term viability of MARP as a village land management tool depends on a number of preconditions, including the following.

An organizational base

The adoption of MARP tools for use in the community planning and management of local natural resources is contingent upon the existence at the local level of an organization capable of providing support for the community in the process of transferring and institutionalizing MARP.

Political will of the support bodies

This new system depends on the willingness of support bodies and government institutions to transfer powers to grassroots organizations. Hence the importance of going beyond mere declarations of intent and of making concerted efforts to devolve power. Where natural resource management is concerned, this should be manifested in a more explicit recognition of the rights of ownership of local communities over the available natural resources.

The existence of community natural resource management programmes

MARP cannot develop from nothing. Its tools are only of value to the community when their use is co-ordinated with development activities already under way and leads to the improved management and strengthened control of resources by local people. Participatory planning and management of local resources is only possible in the context of a coherent development programme. For this reason, while communities may be expected to be able to finance their own development, support from non-governmental organizations and natural resource management projects is still needed. The terms and conditions of partnership should, however, be made very clear, with the external agency relegated to the role of a mere catalyst, so that local potential in the area of participatory management is not stifled.

The implementation of a monitoring system

Mastering MARP as a resource management method is a long process. A monitoring system is therefore essential to ensure that the transfer mechanism guarantees a high level of quality in the use of its tools. Monitoring is perhaps the most critical aspect in the current development of MARP.

CONCLUSION

Despite the potential offered by MARP as a tool in the implementation of a village land management approach, much careful analytical work remains to be done to identify any obstacles to such a link. For example, with regard to training of farmers' organizations, the search for appropriate teaching aids remains a central concern in efforts to facilitate the transfer of tools and techniques. The viability of this method will, however, depend to a large extent on the ability of government institutions to strengthen the powers of local communities in controlling and managing local resources.

4 Supporting Local Natural Resource Management Institutions: Experience Gained and Guiding Principles

YVETTE D. EVERS
International Institute for Environment and Development, London, UK

INTRODUCTION

The subject of natural resource management is a very broad one that potentially covers all the major regions of the world, as well as a range of resources such as land, water, pastures, forests, fish, wildlife and others. What this chapter tries to do is provide a brief synthesis of material describing the experience gained from participatory natural resource management techniques from a developmental perspective. Three case studies highlight different types of management systems and common problems. The chapter also suggests some guiding principles for implementation based on a preliminary review of the lessons of the past.

COMMON PROPERTY RESOURCES AND TRADITIONAL MANAGEMENT SYSTEMS

The field of common property resource management (CPRM) has seen tremendous growth over the last decade and has become the focus of intense interest, particularly by international research and development organizations. Studies in different parts of the world have established the important contribution of local political institutions and indigenous knowledge to the complex resource management and customary land-use regulations that have evolved in dryland areas. These institutions continue to play important roles in both pastoralist and cultivator communities, often complementing or accompanying private management systems. They have also often been misunderstood and their potential contribution neglected (Little *et al.*, 1987).

Natural resources can be held under any one of four property rights regimes: open access, communal property, private property and state property. This

Social Aspects of Sustainable Dryland Management. Edited by Daniel Stiles

chapter will concentrate on the first two types within this analytical typology, although in practice it is more difficult to neatly categorize natural resource management in this way (Murphee, 1993). Any attempt to understand traditional management systems must draw a clear distinction between the commonly confused concepts of open-access regimes (characterized by lack of control and availability of the resources to everyone) and common property regimes (controlled by an identifiable group which regulates access and the rules of behaviour for resource use) (Messerschmidt *et al.*, 1993).

Common property regimes are often controlled or owned by powerful groups (lineages, clans, families, etc.) who can enforce informal regulations and social pressures over a far greater area than is possible under individual ownership. Diminishing autonomy of local institutions over land use, however, makes control more difficult and forces communities to shift to more short-term strategies. In the inland Niger Delta of Mali, for example, effective systems of natural resource management are being destroyed by extractive government policies, the monetization of the Delta's economy and recurrent drought. In the process the linkages between knowledge of resources, dependence on their sustainable production and responsibility for their management are broken down, allowing outsiders with more short-term interests to exploit the local systems and resources.

Photograph 4.1. The inland Niger Delta area of Mali, one of the most important resources in the country, is under threat from lack of effective management systems. (Photo: Daniel Stiles)

The current trend of ineffective state control over resource tenure, and the breakdown of local management institutions, is widespread. However, there are also many examples of highly successful collective management of natural resources. Some represent the continuity of traditional systems, others are formed with external assistance (Makuku, 1993; McCay and Acheson, 1988). In either case, it is increasingly realized that many communities have (or had) effective tenure and natural resource systems that can form the institutional basis for sustainable natural resource management projects. At the local level, the main prerequisite is the establishment or strengthening of such communal property regimes with defined groups that have 'proprietorship' (sanctioned use-rights) over the natural resources concerned, particularly control over exploitation and benefits. At the national level, strong policy objectives will also be required to overcome the reluctance of centralized government authorities to relinquish control over resources (Murphee, 1993).

PARTICIPATORY APPROACHES

Decades of international development efforts have revealed the limitations of a technical approach which places low priority on socio-economic factors in environmental projects. The need for participatory 'bottom-up' approaches that attempt to support local knowledge and management systems is now generally accepted. This is reflected in the recent proliferation of small-scale projects designed and implemented at the village level, mainly supported by NGOs. More recent programmes have attempted an integrated approach of strengthening of natural resource management, such as combining soil and water conservation strategies with household energy issues (Critchley and Graham, 1990).

Participation by the 'beneficiaries' of a given project in its design, implementation and evaluation is a prerequisite for sustainability for several reasons (Toulmin, 1993). First, it is now recognized that indigenous technical knowledge is complex and sophisticated, and can provide a useful basis on which to build interventions. Second, many failures of previous projects can be attributed to their lack of response to local priorities and needs. Finally, establishing local rights and responsibilities constructs a pattern of long-term interests and incentives to engender a sense of 'ownership' of project activities.

Donors must ensure that they do not endorse a weak version of participatory management that does not genuinely allow local communities tenure rights to the resource and its benefits. In other words, collective local management should not be seen as a cheap option for financially hard-pressed government departments (Messerschmidt *et al.*, 1993). A recent review of natural resource management institutions makes it clear that conservation efforts are only undertaken by local people if the benefits outweigh the costs (Murphee, 1993). The validity of this argument is illustrated by the case studies that follow.

CASE STUDIES

While there is an immense body of literature covering both Common Property Resources (CPRs) and the debate on participatory approaches, far less has been published that links these two fields and provides concrete examples of experiences and models that could be used. Despite the rhetoric of concern over environmental degradation and participatory approaches in policy documents over the past decade, it is still early days in terms of the actual implementation of such projects.

The case studies outlined briefly below feature three different types of participatory natural resource management: first, the survival of an existing forest management institution; second, the re-empowerment of a traditional community institution to address land issues; and third, the establishment of new institutions through an externally sponsored wildlife management project. The objective of these case studies is therefore to demonstrate both the range of management institutions and natural resource types that can be found in practice. When analysing such case studies it is also important to draw a distinction between the levels of control exercised over natural resources: no community use allowed (national parks), partial use and responsibility allowed (buffer zones and conservation areas), and authorized local use and management (remaining commons). Often several of these levels of control operate simultaneously in the same area so that the linkages must also be examined carefully.

Case study 1: Survival of traditional management institutions—*Jiri* management, Bikita District, Zimbabwe

With the emergence of the state and central government, especially after independence in 1980, traditional systems that had provided the foundation for sustained management of resources within the territorial boundaries of chieftaincies were widely disrupted. Traditional (informal) leaders lost authority to administer land and resources when government-designed institutions such as the Village Development Committees (VIDCOS) and Ward Development Committees (WARDCOS) were installed. The new formal institutions lacked authority, and as a result formerly communal resources such as forests, grazing and water were rapidly depleted under open-access regimes.

The Norumedzo community in Bikita District of Masvingo Province has been able to continue a long-standing management system of a natural forest, called *jiri*, in their area. An important insect called *harurwa* has been harvested from the *jiri* and bartered for trade for many generations. The origin of the *harurwa* is explained in a myth about the forefathers of the Norumedzo people. A ceremony is organized each year in which the chief selects one of the kraal heads to be in charge of a caretaker team. A rotational selection of representatives from each of the twenty-four villages act as members of this team based at a camp in the forest. Throughout the harvest period they monitor the forest, ensuring that all the rules are abided by. In other times of the year, families can get permission from the chief for the collection of other products such as fuelwood, fruits, caterpillars, and poles as well as water from the indirect catchment effect of the forest.

How does this management system continue in the face of mounting land pressure in the area? Makuku (1993) puts forward three main reasons. First, the community has maintained its traditional administrative structures which are still responsible for regulating resource use and ensuring a sustained harvest from the forest. Second, the *jiri* and *harurwa* are based on strongly held cultural traditions and accepted norms in the community. Access is equitable and each member must contribute to the management, thereby minimizing conflict. Finally, although the traditional authorities have no legal rights to administer the communal resources such as the *jiri*, they have maintained recognition and respect by paying tributes of *harurwa* to neighbouring chiefs, the district administrator and the police. The combination of these factors was important in ensuring that the traditional systems for the management of common natural resources survived.

Case study 2: Re-empowering of traditional management institutions—*Dagashida* assemblies, Bariadi District, Tanzania

Bariadi District lies in Shinyanga Region, formerly the vast territory of the Sukuma chiefdoms. Overgrazing and soil erosion have been perceived as a serious problem by the government and external agencies for many decades in this agro-pastoral area. Numerous efforts to control 'destructive' traditional practices have not met with success. What are some of the factors behind the decline in natural resource management? In some cases traditional institutions (such as the *dagashida*) may need to be adapted to deal with new circumstances.

Participatory research in the area revealed that customary Sukuma land use was characterized by a system of enclosures for the management of natural regeneration which was largely disrupted during the villagization process in the 1970s. It was also found that the traditional institution of *dagashida* had not been operational for nearly twenty years due to imposition of state governance through formal village councils and statutory law.

The *dagashida* is a traditional community assembly which meets to formulate sanctions and 'customary law', and where members of the community who violate these rules are held responsible. Smaller disputes are handled at the neighbourhood level, while twice a year unresolved disputes are dealt with at a large *dagashida* which gathers together men from several villages. The rules of this form of participatory democracy are strictly followed, each age-set having the opportunity to put forward their recommendations for reconciliation. Once the final decision has been made by the elders it can no longer be challenged. How could this form of community organization be harnessed to tackle environmental issues?

At the joint initiative of the research team and several villages a special *dagashida* was held, which also included women for the first time, to identify the conditions that had frustrated spontaneous efforts to enhance natural regeneration and manage non-cultivated land. Several decisions were taken. The *dagashida* should resume its former responsibilities in regulating access to natural resources and, by implication, the district authorities should withdraw to avoid competing systems. Tree tenure should be recognized on farm, grazing and forest land to control use. Sanctions were established to block the use of specified cattle tracks by farmers, and on cattle passing through farmland without prior consent. Rules were established for the collaborative digging and maintenance of shallow wells, and to regulate water access for different user groups. Sanctions were to be imposed on uncontrolled bush fires. Finally, the *dagashida* commissioned further research on indigenous culture and knowledge.

As in the first example, this case study emphasizes the importance of mechanisms for minimizing and resolving disputes. A traditional system was re-empowered, and potentially adapted to assume new powers under the current political conditions. Although the revived *dagashida* unleashed a power struggle in some of the villages, participatory democracy can parallel and complement existing administrative structures, particularly in reinforcing and reconstructing effective common property management (Johansson and Mlenge, 1993).

Case study 3: Establishment of new management institutions—CAMPFIRE Association, Zimbabwe

Although the Communal Areas Management Programme for Indigenous Resources (CAMPFIRE) Association is often referred to in the literature, it is still one of the first attempts at integrating communities into natural resource management and can provide some important lessons.

The 1975 Parks and Wildlife Act conferred proprietorship of wildlife resources on private land. The CAMPFIRE programme aims to extend the positive effects of this policy on communal lands by devolving power and decision making over wildlife and other resources to local people. A high-value resource such as wildlife has successfully generated benefits such as revenue earned through hunting licences, and sales of skins and trophies. Although the focus has been on wildlife, such revenue provides strong incentives to broaden the programme into management of related resources at the local level such as woodlands, grazing and water.

A recent review of CAMPFIRE argues, however, that benefits from communal resources do not necessarily imply that a satisfactory system of management and use which regulates access to these resources has been established—

one which combines production, management, authority and benefit (Murphee, 1993; FTP, 1991). The District Councils were granted rights and responsibilities over the resources and, in theory, to the villages they are elected to represent. The councils are not, however, the producers or managers on the ground and therefore do not represent an effective long-term management system as defined above.

Through the CAMPFIRE process, communities formed new wildlife management committees or wildlife trusts to represent their interests at the district level. Some councils are now taking steps to further delegate proprietorship to the local level with considerable success. This case study demonstrates that a 'demand-driven' need to manage common property resources and balance individual and collective interests can also act as a catalyst for institutional development (Murphee, 1993).

EXPERIENCE GAINED

Keeping these case studies in mind as positive examples of natural resource management approaches, what problems have past projects faced and what lessons can be learned from them?[1] Donor organizations generally do not have a good understanding of the social and economic dynamics of the production systems of the rural societies they work with. Activities (such as wells, schools and clinics) are rarely linked with natural resource management. Such inputs into the community also often confuse participation with better compensation, particularly when resettlement outside of protected areas is involved.

Another common problem is the failure to define clearly who within the community will participate, manage and benefit. A serious lack of institutional analysis to identify effective local community structures that already function within the society has led to a proliferation of institutions created in response to the outside interventions. These new institutions, such as well committees, may undermine existing ones or are too expensive to maintain after project life.

The potential contribution of natural resource management continues to be neglected in many projects. Possible reasons for this neglect include a focus on private property as a basis for rural development, an inadequate understanding of the relative dependence of different households and interest groups on CPRS, and a lack of recognition of the complementary role of common resources to private ones (Jodha, 1985).

A considerable body of 'success stories' now documents more positive experiences and ways to avoid these common problems (Rochette, 1989). The key ingredients for effective projects have been summarized by Chambers and can be usefully incorporated into efforts to support local natural resource management institutions (in Conroy and Litvinoff, 1988):

- A flexible teaming-process allowing projects to change course
- Putting people's priorities first

- Providing secure rights and gains for the poor
- Sustainability through self-help
- Ensuring high staff calibre, commitment and continuity

GUIDING PRINCIPLES

Although the concept of community participation in natural resource management is being more widely accepted, the means by which it can be achieved are not fully understood. How can outside agencies best promote collective management systems? An initial set of guiding principles is suggested below.

1. Support

Existing local informal institutions, however imperfect, should be strengthened. Traditional leadership can use social pressure for compliance to laws and provides mechanisms for minimizing and resolving disputes.

2. Empower

Communities must be granted genuine proprietorship—'the right to use resources, determine the mode of usage, benefit fully from their use, determine the distribution of such benefits and . . . the rules of access' (Murphee, 1993). Such clearly defined rules and regulations for the access, management and equitable sharing of benefits from the common property resource are prerequisites for the achievement of sustainable use.

3. Enable

The further development of collective natural resource management must be favoured by changes in state regulations in respect to tenure on common property and/or co-operative organization. Without such changes, conflicts and power struggles will be inevitable and the management system may not succeed.

KEY QUESTIONS

Apart from the broadest outline of principles such as those suggested above, there is no certain formula for success. Each location and culture will present its own particular issues and problems that require special attention. Some key questions to ask in strengthening or evaluating existing efforts, or designing new participatory natural resource management initiatives are:[2]

- To what extent have groups been identified within the community that could be effectively involved in the management of natural resources?

Photograph 4.2. To effect the three guiding principles for promoting collective management systems the local people need to be listened to. (Photo: Daniel Stiles)

- As defined by the project, what constitutes community involvement in natural resource management?
- What benefits accrue to different members of local communities from their involvement within a meaningful time frame?
- How do the communities themselves view natural resource management?
- What 'bundle of rights' exists on particular resources?
- What government legislation exists regarding natural resource management, and how does this affect community access?
- In what ways do projects or conservation authorities constrain community involvement in natural resource management?

CONCLUSION

Through the use of three case studies this chapter has shown that, in seeking solutions to resource-use problems, there is a need for initiatives that build on the viability and utility of local knowledge and traditional management systems. Experience has shown that the key to successful support of local institutions includes participatory approaches and genuine proprietorship over the natural resources concerned. The challenge remains to apply these principles in the field and tailoring support in a way that will improve the prospects of sustainable development.

NOTES

1. These lessons and problems have been identified by Charles Lane and Richard Moorehead at IIED in their recent work on community conservation projects.
2. Charles Lane (IIED) identified these questions for his work on community conservation projects.

REFERENCES

Anderson, D. and R. Grove (eds) (1987) *Conservation in Africa: People, Policies and Practice*. Cambridge, Cambridge University Press.

Baxter, P.T.W. (ed.) (1991) When the grass is gone: development intervention in African arid lands. *Seminar Proceedings* No. 25. Uppsala, Scandinavian Institute of African Studies.

Bovin, M. and L. Manger (eds) (1990) *Adaptive Strategies in Africa in Arid Lands*. Uppsala, Scandinavian Institute of African Studies.

Bromley, D.W. (ed.) (1992) *Making the Commons Work: Theory, Practice and Policy*. San Francisco, CA, Institute of Contemporary Studies.

Brown, M. and B. Wyckoff-Baird (1992) *Designing Integrated Conservation and Development Projects*. Biodiversity Support Program, WWF, The Nature Conservancy, and World Resources Institute.

Chambers, R., A. Pacey and L. Thrupp (eds) (1989) *Farmer First: Farmer innovation and agricultural research*. London, Intermediate Technology Publications.

Conroy, C. and M. Litvinoff (eds) (1988) *The Greening of Aid: Sustainable Livelihoods in Practice*. London, International Institute for Environment and Development.

Critchley, W. and O. Graham (1990) *Looking after our land—Soil and water conservation in dryland Africa*. London, Oxfam/International Institute for Environment and Development.

FTP (1991) Approaches to wildlife development: lessons from Zambia and Zimbabwe. *Forest, Trees and People Newsletter* No. 13.

Harrison, P. (1987) *The Greening of Africa*. London, Paladin.

Jodha, N.S. (1985) Market forces and the erosion of common property resources. In: *Agricultural Markets in the Semi-Arid Tropics*. Proceedings of an International Workshop. ICRISAT, Patancheru, India.

Johansson, L. and W. Mlenge (1993) Empowering customary community institutions to manage natural resources in Tanzania: Case study from Bariadi District. *Forest, Trees and People Newletter* No. 22.

Johnson, D.H. and D.M. Anderson (eds) (1988) *Ecology of Survival: Case Studies from Northeast African History*. Boulder, CO, Westview Press.

Little, P.D., M.M. Horowitz and A. Engre Nyerges (eds) (1987) *Lands at Risk in the Third World: Local-Level Perspectives*. Boulder, CO, Westview Press.

Makuku, S.J. (1993) Community approaches to common property resources management: The case of Norumedzo community in Bikita, Zimbabwe. *Forest, Trees and People Newsletter* No. 22.

McCay, B. and J.A. Acheson (1988) *The question of the commons: the culture and ecology of communal resources*. Tucson, AZ, University of Arizona Press.

Messerschmidt, D.A. *et al.* (eds) (1993) *Common forest resource management: annotated bibliography of Asia, Africa, and Latin America*. Rome, Food and Agricultural Organization.

Murphee, M.W. (1993) Communities as source management institutions. *Sustainable Agriculture Gatekeeper Series* No. 36. London, International Institute for Environment and Development.

Ostrom, E. (1990) *Governing the Commons: The Evolution of Institutions for Collective Action.* Cambridge, Cambridge University Press.
Rochette, R.M. (1989) *Le Sahel en lutte contre la désertification: Leçons d'expériences.* Weikerheim, CILSS/Verlag.
Toulmin, C. (1993) *Gestion de Terroir: Principles, First Lessons and Implications for Action.* London, IIED Drylands Programme discussion paper prepared for UNSO, International Institute for Environment and Development.

5 Departure Points: Researchers, Rural Communities and the Transfer of Technology

R.D. AYLING
International Development Research Centre, Ottawa, Canada

INTRODUCTION

A little over ten years ago, Canada's International Development Research Centre (IDRC) provided support for the first phase of a project to develop technologies for arid and semi-arid farming in eastern Africa. The project had all the elements for success: a multidisciplinary team of committed national researchers (foresters, agronomists, economists), the participation and technical skills of international scientists, and adequate donor funding which was locally controlled. The project has entered its third and most likely its final phase and provides a useful study of the concept of *technology transfer*.

Farms in the project area vary in size from a few to several hectares; most are divided into croplands, grazing lands and home compounds. Some farmers own or lease draught oxen for cultivation; some raise cattle and most keep goats and poultry. The principal food crops are maize, beans, pigeonpeas, and cowpeas. Most household heads are men while women do the farming with the assistance of their children when not at school.

Many farms have a high proportion of sloping land and their soils are generally infertile and often badly eroded. Annual rainfalls are less than 600 mm and unreliable. There is a long, hot, dry period of several months. Farming under such conditions, to say the least, is a risky occupation.

The overall objective of the project as agreed by the researchers and the donor agency and set out in official documents was: *to develop acceptable agroforestry technologies to improve the quality of life of resource-poor farmers*. Based on an initial survey of the area, the researchers found that the problems associated with dryland farming were: soil erosion, poor soil fertility, and shortages of fodder, particularly at the end of the dry season. The low rainfalls and labour scarcities were considered to be the major constraints to increasing farm production.

Social Aspects of Sustainable Dryland Management. Edited by Daniel Stiles

A number of specific objectives which could be achieved in the short term were proposed. These included studies to evaluate hedgerow intercropping (alley cropping) to improve crop production, species trials for live fences and for fodder and fuelwood production, and the development of a 'technological package' to rehabilitate degraded grazing lands. For the first couple of years, in order to control inputs and obtain quantitative data, work was carried out on the research station and on small, well-fenced plots on land provided by farmers. A number of trials were established, some managed by the researchers, others with farmer support.

There have been a number of technical achievements over the past decade. Several tree species for alley cropping have been identified which can adapt to arid conditions. Biomass production, however, tends to be low with wide variations and higher from station trials than from farm trials. The hedgerows also appear to take up a considerable amount of space in the cropland much to the concern of farmers, and roots and branches sometimes interfere with ploughing operations. Some farmers also complain about the labour required to manage the system.

Other species show potential for use as live fences and fodder banks. However, a major constraint is the labour required for the establishment and protection of the seedlings. Fodder banks also require labour to cut-and-carry branches for livestock feeding. Live fences appear to have limited value without good farmer management to replace missing plants and pruning to increase the density.

For eroded grazing lands, the project team has developed a full rehabilitation package on plots provided by two farmers. The main components include site protection by fencing, the use of micro catchments and pruning to promote faster growth of indigenous trees, ditch construction for erosion control, and the introduction of fodder grasses and tree species. Improvements are clearly visible today and these plots serve as a demonstration to neighbouring farmers. Unfortunately, no farmers have yet adopted these practices, possibly because of the labour required but also perhaps because much of the land to be rehabilitated is communal land and not privately owned. Furthermore, it is now recognized that management efforts are intensive for land which is considered to be low in productivity, i.e. the return on labour is low.

In spite of working closely with a few self-motivated farmers, and in recent years with women's groups and primary schools, widespread community adoption of the agroforestry technologies has not taken place, much to the frustration of the researchers and to the donor agency. The project can best be described, like others of its type, as one of 'limited success'. To be fair, the exercise has been a learning experience to all involved, but it seems to have little to show for so much time and effort. A further indignity was an in-house consultancy report in 1992 which questioned 'the lack of saliency of the agroforestry technologies being offered for adoption'.

Hindsight, always useful, suggests basic issues which require attention. An important factor, unrecognized or not understood by the researchers, is the

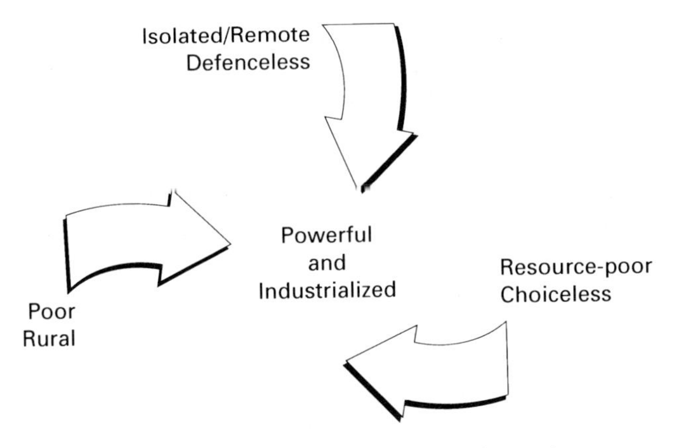

Figure 5.1. The concept of 'cores-and-peripheries'

different 'points of departure' between the farmers and themselves, best explained by Chambers' (1983) concept of 'cores-and-peripheries'.

At the core or centre of the relationship between researchers and farmers are the researchers—knowledgeable, educated, in control, and relatively well-off (Figure 5.1). Most return to urban settings in the evenings. Some, born in rural areas, have been fortunate to escape rural hardship; many have never been farmers, especially subsistence farmers. Researchers have information and access to resources and powerful individuals.

Away from the core, on the peripheries of decision making, are the majority of farmers, with less formal knowledge and power. By definition, they are rural and the more marginal their land, the more vulnerable they are and the greater their poverty. According to Chambers (1983), professional conditioning and training has conditioned the way we see things. Outsiders, non-rural dwellers, whether Third World researchers or First World donor agents, decide on what to do and where to do it.

Recognizing and understanding the different points of departure between rural people and researchers should be a critical first step in project design. It seems obvious but development practitioners need to know what, if anything, rural people want, what their goals and objectives are, and how they make a living—to develop what Diane Rocheleau (1987) has called a *user perspective*, in other words, to attempt to see the issues through the eyes of the farmers. This requires continued examination of assumptions on the part of the researcher, and a fundamental reversal of attitude and perspective.

As an example, the overall objective of this project, 'to develop acceptable agroforestry technologies to improve the quality of life of resource-poor farmers' should be turned on its head to put the people first, and then it becomes: *to work with a specific group of people, poor rural men and women, to help them gain for themselves and their families more of what they want and need through an understanding of the ways in which they make a living.*

Not perfect or complete but a beginning. The focus of attention is on the people, not on the technology. The main objective is to enable people to gain control over their own lives through active participation in their own development. Unfortunately, very few of us are taught at university how to 'empower' rural people, how to find out and understand what their real priorities and problems are, and how to help them explore and develop their potential.

Another early step should have been to recognize the different biophysical and socio-economic conditions, the 'departure points', of researchers and small-scale farmers. Research on research stations is under controlled conditions, on flat land and on fertile soils and usually with unlimited access to water, pesticides, capital equipment and information (Chambers and Jiggins, 1987). Station experiments, if carried out by competent researchers, rarely operate under conditions of risk and uncertainty due to weather, insects, disease, or market fluctuations.

Farmers work under different conditions and therefore the results and observations of different technologies used on research stations and on farmers' fields can be significantly different. In an economic analysis of an alley cropping experiment in West Africa, Arnold (1987) found that maize intercropping gave the highest yields and returns to labour over monocropping but that these results were from on-station trials and might not be the same under farmer-controlled conditions.

Photograph 5.1. Research stations, as here at Katumani near Machakos, usually have more favourable conditions than those found on farms. (Photo: Daniel Stiles)

Photograph 5.2. Subsistence farmers have little control over conditions, and they make do with whatever they have, as here in the IDRC project area. (Photo: Daniel Stiles)

Chambers (1988) refers to this type of farming as 'complex, diverse and risk-prone' or *CDR farming*, complex in activities, diverse in environments, and one where risk reduction is a major preoccupation. Farmers usually depend entirely on family members for labour, and they may own, rent and/or share all or portions of the land they work. Many struggle to meet both consumption and production goals.

If this difference between researchers and farmers were not enough, it has been noted that farmers do not 'just farm'. The raising of crops and animals may not be the most important activity of the family. Income generation is an important objective, income which is often earned from non-farming activities and from off-farm employment (Arnold, 1987). Some of the contact 'farmers' involved in this IDRC-funded agroforestry project were school teachers, one owned a small kiosk, and another worked for the government. An important limitation to projects and programmes designed to 'improve' the circumstances of small-scale farmers it to underestimate the importance of non-farming activities. In East and Southern Africa, additional family income may come from a variety of sources—the making of handicrafts, brewing of beer, trading in local markets, teaching, and wage employment—all of which serve to reduce the time available for farming. Zinyama (1988) found that shortages of family labour on communal farmlands in Zimbabwe was a major constraint to increasing crop production. Many men were away working in urban areas or on large commercial farms, and their wives, the actual 'farmers', had social and family commitments in addition to farming.

Developing and promoting technologies directed only at 'farming' may therefore often prove futile, and have little to do with the quality of the technology. For people involved in a variety of activities to improve their livelihoods, a strategy to support the diversification of rural economies, to increase the opportunities and options available to rural people, might be one worth considering.

It should come as no surprise that the priorities, objectives and needs of rural people can be quite different from those of researchers. A long-overdue socio-economic survey of the rural community involved in the agroforestry project found that the farmers had three main criteria for judging the usefulness of the technologies promoted by the researchers:

- Proof of rapid returns with minimal risk
- Sufficient returns for the resources invested
- The extent and timing of the technologies' demand on scarce resources, especially labour.

This is not the first project or programme nor will it be the last to set off on the development track armed with faulty assumptions and pretensions of knowing what people need.

Strategies to tackle fuelwood scarcities provide some good examples. Wood for cooking, heating and lighting is a basic necessity for many African families. The overcutting of natural woodlands for firewood and/or charcoal production is often regarded as the first step in land degradation and eventual desertification. A two-pronged attack to increase supply and to reduce demand was initiated by most national governments with the support of international donor agencies.

Supplies could be increased by establishing government-controlled fuelwood plantations, some of these on the outskirts of urban centres. It was assumed that the main produce of these plantations would be firewood and hence the poor would directly benefit. However, establishment and maintenance costs are generally high and firewood is either heavily subsidized or is too expensive to purchase. In many cases the plantations are allowed to grow beyond the firewood stage and the wood is then sold as higher-value poles or sawlogs. More critically, the establishment of such plantations often changes the land from local use to one under state control, alienating land that was previously used for communal grazing and the collection of wild foods, firewood and building material. Government plantations thus frequently removed rather than created a valued local resource. Many such initiatives have had the opposite effect of what was intended and led to increased pressures on remaining woodlands (Ayling, 1992).

As these initiatives were questioned, strategists turned to village and communal woodlots. The term 'community forestry' became a popular theme although few proponents understood the concept of 'community' (Cernea,

1990). Nevertheless, it was believed that the required massive planting of fuelwood species could best be accomplished by involving large numbers of people. Communities were expected to mobilize their members to plant and collectively protect the young plantations, and to ensure the fair and equitable distribution of the benefits. However, as labour is invested, the wood has a higher value if markets are available and is unlikely to end up as firewood. Further, although natural woodlands may be used communally with minimum groups rules, the participants in communal woodlots can become suspicious over the sharing of costs and benefits. There have been 'limited successes', in spite of over $160 million spent on various community forestry programmes between 1975 and 1982 (Barrow, 1991). Money has not been a constraint.

At the same time, the demand for fuelwood could be reduced by developing more efficient woodstoves. Technically, the products of most stove programmes were excellent. They were more energy-efficient than the traditional three-stone fire and saved fuel and money. Many were made from locally available materials such as clay or brick; some were portable and made of metal lined with clay. Most could be produced inexpensively by local craftsmen. However, according to Shepherd (1990), many stove programmes failed to have an impact. In some instances, the stoves were used inefficiently; their distribution was often limited, and in some cases, the stoves were relatively expensive. More importantly, although many were technically sound, many were simply culturally unacceptable.

Researchers frequently underestimated the social importance of the traditional fire. Families involved in the pilot testing of 'improved' stoves were often also using the more sociable open fire, especially for light and warmth. Researchers also failed to consider that in real life, whenever fuel is scarce or costs increase, people respond in a variety of ways to reduce consumption (Deewees, 1989). For people of developing countries, this includes better fire tending and management, different ways to prepare food using less energy, the use of other fuels such as agricultural wastes, and when possible, more time spent collecting firewood. This should not be surprising. It is a universal response to scarcities South and North—using less more efficiently, using alternatives, changing life styles.

There are other examples of 'limited-success' technologies developed by outsiders although not many appear in the literature. Africa is littered with the carcasses of abandoned quick fixes—neglected species trials, over-mature communal woodlots, solar dryers, improved cookstoves, wind-powered generators, testimony to good intentions gone bad.

The local situation in rural communities is complex and changing, with accelerated rates of change having direct effects on already fragile social systems and environments. It is therefore surprising, not to say arrogant, that many development workers think they have the right answers and easy solutions to these complex problems (Meyeroff, 1991). In many instances, they do not even have the right questions. Barrow (1991) suggests, for example, that almost all development interventions have not helped impoverished pastoralists because of an inadequate understanding of pastoralism. Pastoralists, if they survive, do so despite 'development projects', not because of them.

Rural people are the experts in rural living and survival. They should be at the core of development but often have too little involvement in planning and design. Their first opportunity often comes when development workers show up on their doorstep, long after their collaboration has been taken for granted and research topics decided.

Poverty and ecological degradation are interrelated. Poor people do overexploit their resources out of necessity, sacrificing their future to survive the present. Ecological deterioration perpetuates poverty as degraded ecosystems offer diminishing returns (Durning, 1990). Efforts focusing on environmental degradation while ignoring the root causes of poverty will have mixed results. Simplistic explanations and solutions will have little positive effect. The popular theme of *technology transfer* is flawed, based on core beliefs and assumptions of what will work and on what is best for rural people. Successful aid, however, builds on locally generated efforts to rehabilitate drylands. Development workers have to make the effort to talk and to listen to rural people, to discover what their problems are and what solutions they think are possible. As Durning (1990) observes, 'for development to help the poor, it must put them first . . . true development does not simply provide for the needy, it enables them to provide for themselves'. Development must start and build on what people know. Only then will it be sustainable and result in self-reliance rather than dependency.

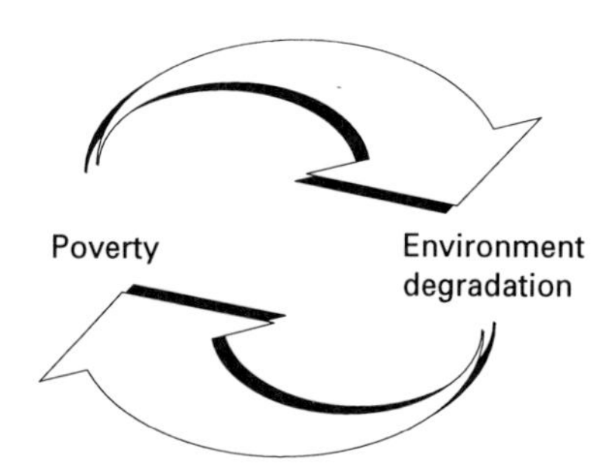

For every problem there is a solution
that is simple, direct and . . . wrong

H.L. Menken

Figure 5.2. Poor people overexploit their resources out of necessity

REFERENCES

Arnold, J.E.M. (1987) Economic considerations in agroforestry. In Stepler, H.A. and Nair, P.K.R. (eds), *Agroforestry—A Decade of Development*. Nairobi, International Centre for Research on Agroforestry, pp. 173–90.

Ayling, R.D. (1992) Changing perspectives of the demands on tropical forestry. *The Forestry Chronicle*, **68**(4), 476–80.

Barrow, E.G.C. (1991) The challenge for social forestry extension work in pastoral Africa. *ODI Social Forestry Network Paper*. 12e.

Cernea, M.M. (1990) Beyond community woodlots: programmes with participation. *ODI Social Forestry Network Paper*. 11e.

Chambers R. (1983) *Rural Development: Putting the Last First*. Harlow, Longman Scientific & Technical.

Chambers, R. (1988) Sustainable rural livelihoods: a key strategy for people. In: Conroy, C. and M. Litvinoff (eds), *The Greening of Aid—Sustainable Livelihoods in Practice*. London, Earthscan Publications, pp. 1–17.

Chambers, R. and J. Jiggins (1987) Agricultural research for resource-poor farmers. I. Transfer-of-technology and farming systems research. *Agricultural Administration & Extension*, **27**, 35–52.

Dewees, P.A. (1989) The woodfuel crisis reconsidered: observations on the dynamics of abundance and scarcity. *World Development* **17**(8), 1159–72.

Durning, A.P. (1990) Ending poverty. In: *State of the World 1990*. Washington, DC. Worldwatch Institute, pp. 135–53.

Meyerhof, E. (1991) *Taking Stock: Changing Livelihoods in an Agropastoral Community*. Nairobi, African Centre for Technology Studies.

Rocheleau, D.E. (1987) The user perspective and the agroforestry research and action agenda. In: Gholz, H.L. (ed.), *Agroforestry: Realities, Possibilities and Potentials*. Dordrecht, Martinus Nijhoof Publications, pp. 59–87.

Shepherd, G. (1990) Forestry, social forestry, fuelwood and the environment: a tour of the horizon. *ODI Social Forestry Network Paper*. 11a.

Zinyama, L.M. (1988) Farmers' perspectives of the constraints against increased crop production in the subsistence communal farming sector of Zimbabwe. *Agricultural Administration & Extension* **29**, 97–110.

6 Rajasthan's Camel Pastoralists and NGOs: the View from the Bottom

ILSE KÖHLER-ROLLEFSON
League for Pastoral Peoples, Germany

The futility of top-down approaches when implementing dryland management projects—or any grassroots level project—has gained widespread recognition. But are donor agencies and NGOs really able to change their ways and act in response to problems *as they are perceived by local populations?* Experiences with setting up a project 'at the request' of Raika camel pastoralists in Rajasthan, India, suggest that fundamental changes in the way developing agencies operate are necessary before a true 'listening to the people' strategy can be adopted.

INTRODUCTION: THE MACRO-CONTEXT

Rajasthan, India's most western state, encompasses a succession of arid, semiarid, and sub-humid ecological zones from west to east. The Thar Desert that straddles India's border with Pakistan is often described as the world's most densely populated desert. This is certainly testimony to the wisdom of its inhabitants in making thrifty and efficient use of its resources. However, with a population growth rate considerably above the average for India—the population of its arid districts has almost trebled since the beginning of the century—Rajasthan's ability to accommodate painlessly a further increase of its human and animal populations may be close to exhaustion. This is indicated by the bitter atmosphere that has developed between social groups whose economic activities once complemented each other. Peaceful coexistence is replaced by an increasingly belligerent struggle for land, or access to it, between agriculturalists and pastoralists and among animal herders belonging to different castes. In an unprecedented event nine shepherds belonging to the Raika caste lost their lives during the 1993 rainy season after being ambushed by over one thousand members of a different caste in a pasture conflict.

Rajasthan's predominant land-use strategy until India's independence (1949) was sheep, camel, and cattle pastoralism that relied on communally

Social Aspects of Sustainable Dryland Management. Edited by Daniel Stiles

Photograph 6.1. With population increase in Rajasthan since the 1950s, livestock density has trebled. (Photo: Daniel Stiles)

owned property resources, combined with opportunistic rain-fed cultivation of sorghum and pulses. In the 1950s a more equitable redistribution of the resources, held until then by feudal overlords, was carried out. In the course of these land reforms some previous pasture land was allotted for cultivation. Other developments, including the spread of irrigation agriculture, increased availability of tractors, shortening of the fallowing period, mining activities, and the creation of nature reserves, have also cut into the amount of land available for grazing (Jodha, 1988). Thus, between 1956 and 1987 common property resources decreased by 32%, while the net sown area increased from 28.6% in 1951 to 47% in 1981. Concurrent with human population growth, livestock populations have also risen, and the stocking density has trebled (Centre for Science and Environment, 1985). The effect of these developments on the plant biomass can best be illustrated by quoting Rajasthani villagers, who frequently make comments such as 'earlier there used to be a jungle here, now there is nothing'.

Rajasthan's pastoralists have found a solution to the lack of grazing in their home state by making seasonal moves into adjoining states—notably Madhya Pradesh, where uncultivated land is available, and Haryana, where the post-harvest stubble of irrigated fields can be exploited (Kavoori, n.d.). Largely because of these newly found outlets, substantial profits can still be made from sheep breeding in good years; therefore even members of the landowning castes have fairly recently taken to migratory livestock production. Yet, for the traditional pastoralists who own no land and depend entirely on common property resources, the situation is becoming more and more difficult (Köhler-Rollefson, 1992).

No disinterested onlooker would dispute the need to reverse the present trend of increasing animal numbers and the necessity of reaching a long-term equilibrium between available fodder resources and numbers of animals if irreversible damage to the ecosystem is to be prevented. In the words of an eminent local activist, Kishore Saint, 'Even though the comparison with the African Sahelian situation may not be strictly valid, there is no doubt that similar processes of degradation are at work in the subcontinent'.[1] One would therefore expect that the situation of Rajasthani pastoralists would be an issue of prime concern to the large array of international donor-NGOs represented in India. Yet, as will be demonstrated, these donor-NGOs are in a poor position to tackle problems of such a nature because of bureaucratic constraints and the lack of articulation with the actual people.

LISTENING TO THE PEOPLE

The Raika camel breeders

About three years ago I embarked on an academic research project[2] with the mission to understand the Indian camel pastoral system, a subject about which very little was known and nothing had been written. Because the cultural and economic context of camel breeding is entirely different in India from other countries, the most salient features will be summarized here. In Rajasthan, camel breeding is the traditional domain of a particular caste called the Raikas or Rebaris. These are not nomads in the traditional sense, since they have permanent dwellings. They represent one of many specialized professions that compose the interdependent village caste community (Srivastava, 1991). Their camel breeding is not subsistence but market oriented. The Raikas have large herds of camels, but they do not exploit them for food. Killing or slaughtering camels for meat is an absolute taboo, and the milk is only used sporadically. The sole rationale of the system is to provide male camels for sale to rural and urban users. Camel-drawn carts are an indispensable component of Rajasthan's infrastructure, often the only source of communication and transportation in its remoter areas but also ubiquitous in the larger cities. They are also used for ploughing, threshing, lifting water, etc., and there is no doubt that they save India enormous sums of foreign currency that would have to be spent on fuel imports if these tasks were carried out by machines. Hence the demand for camels is strong and there is no indication that it will abate in the near future. It is somewhat ironic that the traditional pastoralists producing these camels are in dire straits.

Encounter with a community in distress

The Raikas have a reputation of being one of the most conservative groups in Rajasthan and as being extremely tight-lipped with outsiders, which poses a severe hindrance to data collection on camel pastoralism. I was lucky enough to enlist as interpreter the first member of this group who had trained as a veterinarian, and it was only due to this circumstance that I could get access to the community and start collecting data. In the course of our enquiries we learnt about a Raika *dhani* (settlement) at the foot of the Aravalli Hills that owned 2000–3000 camels, all expected to congregate there at a certain spring holiday for their annual shearing. We went to the *dhani* on the particular date and indeed encountered a couple of thousand camels, as well as a community which was totally camel oriented, as evidenced by a collection of memorial stelae depicting camel-mounted ancestors and self-made clay camels the children played with. Not surprisingly, I was totally fascinated and felt myself in camel heaven.

Unfortunately, the enchantment was not mutual. While I was used by then to the not overtly welcoming demeanour of the Raikas, in this case they were almost openly rude. When I finally got my assistant to—very reluctantly—translate their comments, the message was 'What is the good of this woman coming here, asking questions, and taking photographs? We have a lot of problems and nobody is doing anything for us.' Having been baited by this conversational gambit, we obviously felt compelled to enquire about the particular nature of their problems which only emerged after a lot of prodding.

The central government had abruptly closed the traditional summer grazing grounds of these Raikas in the Aravalli Hills and the absence of alternative pasture opportunities had brought their herds to the brink of starvation. Probably because of this predisposing factor, their camels had also suffered an especially severe outbreak of trypanosomiasis, that resulted in a massive spate of abortions. The medicines for the treatment of this disease were almost impossible to obtain and in any case too expensive. The people felt that they were facing economic ruin; they were desperate for veterinary medicines for their camels and permission to graze in the forests. Trying to wheedle detailed information out of them proved futile, beyond eliciting bitter complaints. But in order to gauge the extent of the problem, we needed at least a rough estimate of the number of their female breeding camels and the proportion of diseased animals. Only by making the unwise move of promising that we would bring their crisis to the attention of the appropriate authorities and try to find help for their veterinary problems, were we able to establish that around 50% of their pregnant camels had aborted.

My research assistant and I then proceeded to Delhi and presented the case to a selection of donor-NGOs, who were all sympathetic, thrilled to hear for the first time about camels, willing to help and even to part with considerable chunks of money, but all their offers of support were tied to the existence of a registered voluntary organization fulfilling a number of bureaucratic requirements that they could channel their money through.

Since our remote village was not in the catchment area of an already existing voluntary organization, we had only two options: either to abandon the efforts or to engage in the rather paternalistic attempt of persuading our people of the need to organize themselves into a formal group according to the various bureaucratic requirements stipulated by donor-agencies.

To cut a long story short, after over two years of running back and forth between the 'target-community' and donor-NGOs, we still have not succeeded in establishing a link between them. We have not been able to fulfil the formal prerequisites that would enable our community to receive any kind of donor-NGO help and support. In regard to their concepts of what constitutes a worthwhile project and to forms and standards of social interaction, the worlds of the donor-NGOs and the people that need help are separated by a gulf so vast that 'never the twain shall meet'. I must add that not all the 'fault' for this rests with the NGOs. The situation is to some extent self-inflicted by the Raikas who cry out for help, but whose lack in social skills—as perceived by outsiders to their community—makes this difficult.

We had two cases in which representatives from Delhi-based NGOs expressed the intention to visit the 'project area', but there was nobody on the Raika side to answer their letters or receive them. One could put the blame for this at the feet of the Raikas and ask, as one hurried foreign NGO representative once did, 'why help them, if they are like that?' Admittedly, in my more frustrated moments I am inclined to agree with him, but on the other hand, how can the Raikas be expected to open up immediately to outsiders who drive up in a taxi, don't speak their language, are absolutely ignorant about even the most basic aspects of their life and leave after a few hours of asking impertinent questions?

However, while we failed on the front of fulfilling the prerequisites for direct interaction between donor-NGOs and our people[3] there have been other significant developments. The biggest one is that, against the odds, and although we never delivered on the promises that started the whole process, the relationship with the Raikas has markedly improved and people no longer doubt our intentions. Several small steps were successful in creating a positive atmosphere, and I no longer dread the moment I have to leave my vehicle and face the village crowds. A major breakthrough was achieved by purchasing a camel and its subsequent placement in the care of the Raikas. The need to check up on the well-being of this animal provided a legitimate context for visits and enquiries that then were no longer regarded as snooping by an outsider.

The distribution of a limited quantity of medicines against trypanosomiasis, financed by a private fund-raising drive and, recently, the German Embassy in New Delhi, were major public relations coups, and at the same time we

Photograph 6.2. The distribution of camel medicines and the provision of veterinary care helped to gain the trust and co-operation of the Raika. (Photo: Ilse Köhler-Rollefson)

obtained information on herd sizes and rate of losses. In March 1993 we organized an international workshop[4] which brought together the various factions concerned with pastoral issues on an administrative, scientific, and development level. Also participating was a large contingent of Raikas who took the opportunity to present their demands vociferously. While no concrete recommendations emanated from this workshop, which frequently deteriorated into shouting matches, it succeeded in bringing the subject much more into the limelight and raised the level of awareness of the critical situation facing pastoralists in the appropriate circles.

All this may seem hardly worth mentioning, but we feel the steps taken so far have now prepared the ground for successful interaction with the Raikas. Luckily, a newly established and dynamic research-oriented local NGO in Jodhpur, the School of Desert Sciences, has become very interested in the problem and, although located 200 km from the village in question, it will act as the official implementing body for the project we are planning. This whole process has taken up an inordinate amount of time, but it has also given us a chance to reflect on interventions that could contribute to a *long-term* solution. It has dawned on us that the free provision of veterinary medicines as envisioned by the Raikas is not going to achieve this and that more comprehensive measures are required.

THE ACTION PLAN

1. Data-collection and liaison

One of the most frustrating aspects of working with the Raikas is that they desperately want help, but at the same time are reticent to provide even the most basic bits of information. But no support project can be mounted without a solid database and an identification of the bottlenecks in the present camel production system. The Raikas therefore have to be persuaded to cooperate on this level. We think that they will change their attitude if they realize that backing up their complaints with accurate figures on the amount of losses or numbers of abortions their camels have suffered will put them into a much stronger position *vis-à-vis* government agencies. We have had surprisingly good success with employing a young literate Raika from a camel-breeding family for the sole purpose of systematic data collection on aspects that are essential for outsiders to understand the Raika camel production system. He is doing this *at his own pace and according to his own judgement*, but has already accumulated an extensive body of data on kinship patterns, herd sizes, herd composition, migratory patterns and so forth.

Our hope is that he will eventually become the essential link between his community and the donor community, and grow into a liaison-person who is capable of presenting the worries and viewpoints of the Raikas to outside agencies. For this reason we are also financing his English studies and gently familiarizing him with the NGO scene.

2. Camel Husbandry Improvement Project (CHIP)

I mentioned in the beginning that in Rajasthan there is no leeway for further expansion of livestock numbers. The camel holdings of the people in our village have already drastically been reduced in the past decades from 10 000 head, or 'so many that we did not even care if we lost one', to the present 2000–3000. Yet from the outsider's perspective (who is, of course, not familiar with all the intricacies of the situation), it seems as if a further reduction of animal numbers is desirable in view of the fact that the Raika camels are chronically undernourished. This would not necessarily entail a reduction of income. It is conceivable that a smaller number of well-fed camels with a low incidence of disease and hence higher reproductive rate would generate the same or more profits as the present large number of camels that teeter at the brink of starvation. It just *might* be possible that by raising inputs in terms of feed and veterinary care, the efficiency of the present low-input system could be improved.

Because there is no guarantee that a higher-input strategy would be successful, it obviously cannot be imposed on the Raikas. But the approach is well worth testing by establishing a couple of experimental camel herds kept under different management conditions and with varying levels of input in veterinary care and fodder provision. The productivity and cost/benefit ratios of these herds can then be compared with those kept under traditional conditions. The intention is to conduct this applied research under the close scrutiny of the Raikas and to provide them with an opportunity to make suggestions and act in an advisory function. The outcome cannot be predicted, but at least a mutual learning process about camel production would be set in motion that could eventually lead to a successful amalgamation of traditional knowledge with 'modern science'.[5]

3. Pastoral Information Centre (PIC)

The efforts at the village level have to be combined with a regional approach in order to make more than a strictly localized impact. But there is a tremendous degree of variation in pastoral adaptations within Rajasthan, and a solution for the problems of the people of our village in its fertile agricultural setting would not be applicable to camel breeders in the far west who operate in an entirely different ecological context. Again, lack of information on the strategies and migratory patterns of pastoralists and their perceptions about their problems is the first obstacle to overcome. For this purpose, the School of Desert Sciences and the League for Pastoral Peoples are planning to establish an information centre that will systematically compile information on these and all related aspects, such as legal aspects of land use and tenure. Data are to be collected through our own surveys, through networking with researchers and local institutions, and by acting as a depository for all published and grey literature on pastoralism. Besides this official purpose, the centre—which we are hoping to establish in Jodhpur—will have the additional but equally important function of acting as a forum for communication between pastoralists and people who interact with them on a research, administrative or development level. We envision PIC to function as an informal meeting place with a hospitable atmosphere where pastoralists can drop in on their sporadic visits into town, pick up information, as well as report about special incidents, air grievances, and even lodge formal complaints.

SUGGESTIONS FOR IMPROVEMENT

From the perspective of somebody who is trying to solve a particular problem encountered in the field, I would like to venture the following suggestions to donor-NGOs to improve their articulation with the people. Some of them may be particular to the Indian context and NGO set-up there, but others are of general relevance.

Country-based donor NGOs

The donor-NGOs in Delhi are, rightly, concerned with spending their money wisely and to this end they support only well-established voluntary organizations. These are generally run by outsiders, i.e. non-Rajasthani natives, who possess charisma and who produce impressive reports listing the number of handpumps installed, bore wells sunk, smokeless stoves distributed and opium addicts treated. It is absolutely not my intention to detract from such achievements; certainly these organizations do excellent work. But there are two problems with them.

First, their sphere of influence is locally restricted and they thus cover only small sections of the country. Communities who are situated outside their catchment area such as our village have no chance of receiving any benefits. In order to spend their money responsibly, donor-NGOs queue up to support the limited number of reputable field-NGOs, but they have no mechanisms for reaching the rest of the population. The local availability of reputable field-NGOs thus constitutes a bottleneck in the system which prevents access to the majority of the people. Donor-NGOs are not to be blamed for this, but they could be slightly more adventurous in supporting fledgling organizations and new initiatives.

Second, the established field organizations tend to take on shades of 'mini-empires', partly because they control so much money. It seems (from observation of a limited sample) that they sometimes undertake no efforts to let the deprived communities they are serving eventually become independent of their tutelage. Rather, they seem to keep them in a perpetual state of dependence, showering them with handpumps, energy-efficient stoves, and other gadgets. Again, there is nothing wrong with such interventions, but rather than existing as ends in themselves they should be the starting points for inspiring local people to actively pursue their own development. Ideally, field-NGOs should be able to form satellite organizations that can replicate their efforts in another spot to achieve a broader coverage. I think it is within the power of donor-NGOs to make stipulations that all organizations they support make efforts to that effect.

Donor-NGOs based in the Western world

Foreign NGOs have a different perspective. Being remoter from the field, representatives of NGOs based in Europe are quite keen on the idea of supporting true grassroots efforts. They may even subscribe to the belief that this is what they are doing when they pour money into the established field-NGOs, whom they only know from fleeting visits at annual or bi-annual intervals. Having to channel their money through country-based partner organizations, they are blissfully unaware of the almost insurmountable obstacles a true grassroots initiative would face. Perhaps they cannot be blamed for this, but rather disturbing is the poor opinion many of them have of pastoralists. For them, the research of the last few years that re-evaluated many of the earlier fallacies about pastoralists, such as the 'tragedy of the commons', might never have happened. They subscribe to totally outdated concepts of pastoralists as unproductive and ecologically destructive and would like all of them to turn into settled agriculturalists.

Pastoralists are in dire need of some positive publicity and it is the responsibility of dryland and pastoral development experts such as are assembled at this workshop to remedy this. (The League for Pastoral Peoples was founded, in part, to generate some positive publicity for pastoralists.) Pastoralists have so many positive aspects to them that this poses no difficulty. The indigenous inhabitants of the tropical rainforests have gained recognition as guardians of this ecological zone even among the general public, and there is no reason why pastoralists should not gain the same credit as stewards of the world's drylands. Their role in preserving the genetic animal resources that are the prerequisite for continued exploitation of this zone for food and energy production (Köhler-Rollefson, 1992, 1993) needs to be acknowledged.

From my limited experiences with NGOs it seems as if they are basically not demand-driven and have insufficient capacity to respond to problems as they are perceived by the people and to act spontaneously on crises that arise, except earthquakes and other major natural disasters. It is not that NGOs are not interested or open, but their actions are limited by too many bureaucratic and policy constraints. They also have a tendency to follow very narrowly defined programmes and rigidly push particular concepts and project approaches that they have identified as promising, to the exclusion of any other ideas. They should be more adventurous in adopting innovative approaches and in providing flexible funding schedules.

The 'listening to the people' approach

This term sums up the essence of what has been missing in many development projects. Yet 'listening to the people' comes with its own set of problems, for the needs of the particular people one listens to almost invariably conflict with those of other equally unfortunate groups. The demands of the people one listens to therefore have to be balanced against those of other communities or segments thereof, as well as against those of the community at large, i.e. the state or even the rest of humankind. At least in a situation such as in Rajasthan where resources are extremely scarce, there are so many different factions with aspirations to the same resources that supporting the wishes of one interest group will almost out of necessity proceed at the expense of others.

Take our example: If you listen to the Raika men as I have been doing, they want nothing dearer than the Aravalli Hills to be opened for grazing and to have an abundant supply of medicines to treat their camels. I sympathize very much with their requests, especially in regard to the Aravalli Hills which they have used for centuries and to which they thus have strong customary rights. Yet if unlimited access to the now-restricted forest areas were granted, this would only be a stop-gap measure with consequences for the habitat of several forest-dwelling tribal groups such as the Bhil and Grassias, as well as wildlife species. Saying this, I do not want to deny the Raikas the ability to manage their resource base, but in the current situation the mechanisms that once guaranteed sustainable exploitation and amicable sharing of resources appear to be no longer functioning. Their impoverished situation and lack of alternatives will *force* the Raikas to overexploit these resources to make ends meet now, even if they are fully aware that they are mortgaging their future.

The desire for an unlimited supply of veterinary medicines is a charitable cause in its own right. But this would be financially unsustainable taking into account the current relationship between the costs of medicines and camel prices. Indiscriminate use of trypanocides could also lead to drug resistance of the microorganism that causes the disease. Hence for a solution to the problem, other less convenient as well as less costly disease-control strategies will have to be explored, even if the need for this is not immediately obvious to the Raikas, which will test their patience.

If you ask Raika women about their perceptions of the problems that face their community, they will rarely refer to matters of animal production. Instead, they will tell you that opium consumption and the large amounts of money spent by men on death feasts are the predicaments at least partially responsible for the demise of their community. While I wholeheartedly agree with them, this is a matter I would be far too timid to raise in the presence of Raika men. It would jeopardize my present positive but precarious relationship with the community and will have to be borne in mind for future reference.

But the requests of the Raikas also conflict with those of other, even more deprived, communities. One example concerns the exotic Acacia species *Prosopis juliflora*, imported and propagated during the British occupation as a windbreak and source of fuel. The Raikas would like this plant to be eradicated, because it is useless as animal fodder and its thorns cause damage to their animals' feet. I used to faithfully echo their sentiments, until one day I met a group of itinerant yogis, i.e. snake charmers, who extolled the virtues of this shrub as raw material for building shelters. They were totally opposed to the eradication programmes sponsored by some NGOs.

'Listening to the people' thus endows the listener with a perturbing responsibility and it invariably puts him or her into a very troublesome paternalistic position, for he or she then has to weigh all the different pros and cons, set priorities and even decide whose needs are most urgent, whose requests to leave for later, and whose to ignore. Although only the rural people can judge the impact of interventions at the local level, the wider implications of particular development strategies can, at present, be evaluated only by those outside the affected communities, since only they are aware of regional and global trends. I think—perhaps I am over-optimistic—that it is possible to share that responsibility with the people at the grassroots level.

At the moment we have a situation where the development worker has an unlimited amount of information at hand, through access to data-banks, e-mail connections to fellow researchers and so on. Compare that with the situation of the pastoralists who rely only on oral communication networks, and are thus cut off from all but the most localized events and those that affect their own community. Thus the Raikas might sense, but are not expressly aware of, the factors that are responsible for their economic decline, such as the population explosion, the fact that livestock density has trebled, and that the CPRs have decreased by x%. Yet there is no reason why, with the help of appropriate educational material, they could not be made aware of these circumstances and learn about the developments in the world at large. By explicitly informing them about the situation in the world at large, the people at the bottom would be given an opportunity to act and react accordingly rather than engaging in the endless and fruitless lamentations over their situation that are only reflections of their helplessness.

Putting some effort into educating people is also appropriate for the sake of achieving a more balanced and equal relationship between the 'people' and the 'outsiders'. Traditional knowledge and 'Indigenous Knowledge Systems' have become one of the most fashionable topics of research, and this angle of enquiry has contributed much to fostering the 'listening to the people' approach. But it is not free from a certain touch of intellectual appropriation. If we extract this kind of valuable knowledge from indigenous people to incorporate it into our electronic data-banks (and often for the purpose of furthering our own careers), asking them to share with us what represents one of the few—even only—assets they have been left with, then we owe it to them to reciprocate in kind and impart relevant parts of our knowledge as well.

Sustainable exploitation of the world's drylands will only be possible if we merge local traditional knowledge with modern science, nuture an atmosphere of mutual learning between the people and the outsiders, and combine awareness of global trends and needs with consideration for local cultural idiosyncrasies.

NOTES

1. 'Desertification and drought in South Asia' in the *Indian Express*, 11 July 1991.
2. Initially research was funded by the American Institute of Indian Studies, followed by a research grant from the National Geographic Society.
3. A voluntary organization has been created, at least on paper, that has as its purpose the improvement of the Raikas' welfare. However, its officers are so called 'educated' Raikas who have already distanced themselves to some extent from their pastoral traditions and believe the purpose should be the upliftment of their community through scholarships for education at various levels. The activities of this organization will probably not be of immediate benefit to Raikas still pursuing a traditional pastoral livelihood.
4. The proceedings of this workshop are expected to appear at the beginning of 1994. The workshop was funded by NORAD, Intercooperation, the Ford Foundation and Misereor.
5. There are several potential ways for improving the profitability of the present system. One of them is looking for additional marketing opportunities for camel products. Presently camels are sold exclusively as draught animals. Camel wool, cheese, and riding camels are other options, but might be unacceptable to the Raikas. Camel meat can be ruled out, since the slaughter of camels is taboo to the Raikas, and consumption of their meat would be violently opposed.

REFERENCES

Centre for Science and Environment (1985) *The State of India's Environment 1984–85. The Second Citizens' Report*. New Delhi.

Jodha, N.S. (1988) The decline of common property resources in Rajasthan, India. *Pastoral Development Network Paper* 22c. London, Overseas Development Institute.

Kavoori, P.S. (n.d.) *Pastoral Transhumance in Western Rajasthan, India. A report on the migratory system of sheep*. Report to the Norwegian Agency for Development, New Delhi.

Köhler-Rollefson, I. (1992) The Raika dromedary breeders of Rajasthan: a pastoral system in crisis. *Nomadic Peoples* 30: 74–83.

Köhler-Rollefson, I. (1993) Social factors in the development of indigenous livestock breeds. Paper presented at the *Pithecanthropus Centennial*, 26 June–1 July 1993, Leiden.

Srivastava, V.K. (1991) Who are the Raikas/Rabaris? *Man in India* 71(1): 279–304.

Part III

SOCIAL ASPECTS OF DRYLAND MANAGEMENT

The chapters in this part present examples of some of the new ideas related to land and natural resource management in the drylands that have resulted from detailed field research and experience. The direct cause-and-effect linkage between population pressure and land degradation has been demonstrated to be too simplistic. Factors that appear to be more important than simple human numbers for land degradation are the way in which people *use* their land and resources, the incentives and possibilities they have at their disposal for making inputs, and their ability and freedom to make decisions regarding resource management.

The approaches and experiences show variability, however. Which approach is best depends on the specific contextual situation, related to subsistence system, settlement pattern (i.e. nomadic, nuclear village, scattered homesteads, etc.), environment, land tenure system, and so on. For example, the Village Land Management approach, which is highly structured, seems more applicable to nucleated villages that have well-defined 'terroir' or village lands. The Machakos example is more diffuse and involves self-help groups introducing new technologies from outside and adapting them to scattered homesteads in hilly terrain. The rangeland management practices introduced from outside have not worked, and it appears that traditional ways are better suited to fluctuating dryland ecosystems.

7 Sustainable Growth in Machakos, Kenya[1]

MICHAEL MORTIMORE, MARY TIFFEN AND FRANCIS GICHUKI
Overseas Development Institute, London, UK

> The Machakos Reserve is an appalling example of a large area of land which has been subjected to uncoordinated and practically uncontrolled development by natives. Every phase of misuse of land is vividly and poignantly displayed in this Reserve, the inhabitants of which are rapidly drifting to a state of hopeless and miserable poverty and their land to a parching desert of rocks, stones and sand (Colin Maher, senior soil conservation officer, 1937).

At that time, the population of the Ukamba Reserve was about 250 000. Extensive livestock raising was combined with shifting cultivation on small hand-cultivated plots of maize and other food crops. Frequent and unpredictable droughts decimated food production and damaged the heavily grazed rangeland. Much natural woodland had been removed and replaced by sparse shrub- and grassland. Farm yields were low and thought to be declining, soil nutrients were depleted, the topsoil was eroding away and livestock numbers were considered to be far in excess of carrying capacity. The official view, as the quotation above makes clear, was that the farming system was unsustainable, if not in terminal decline.

THE MACHAKOS MIRACLE

In 1990 the population of Machakos District had increased by a factor of six to nearly 1 500 000. Although the district had roughly doubled in size, with the accession of previously uninhabited Crown lands, population densities increased from under 80 per km in the wettest areas in 1932 to nearly 400 per km in 1989, and from about 50 per km in the drier areas to nearly 150 per km. The annual rate of population increase peaked at 3.76% in the period 1969–79. Livestock population also increased.

Agricultural output (food and cash crops, horticulture and livestock) increased from less than 0.4 tons per capita in 1932 (converting to maize equivalent at 1957 prices) to nearly 1.2 tons per capita in 1989, and from 10 to

Social Aspects of Sustainable Dryland Management. Edited by Daniel Stiles

Photograph 7.1(a). Eroded country near Machakos town in 1937. Irregular patches of bare soil are visible in the middle distance and gullied, treeless slopes in the far distance. (Photo: Kenya National Archives)

110 tons per km^2 (Figure 7.1). Cash crops, particularly coffee, made up much of the increase during the 1970s, and fruit and vegetable production in the 1980s. Cash crops have only very occasionally occupied as much as 15% of cropped areas as there is a need to keep most land under food crops to avoid purchases at high prices in poor seasons. Nevertheless, because of their higher unit values cash crops are vitally important as income generators. District food sufficiency improved substantially, some households buying, others selling. Living standards also improved through higher cash income.

Soil and water conservation structures were extended, during 1960–90, to almost 100% of the district's arable land, except only the flattest and least densely populated areas. By 1990, rangeland was also coming under increasingly careful management. Planting and protecting trees on smallholdings became universal practice. Measured tree densities were found to be highest on the smallest of holdings. The farming system was more, not less, sustainable than thirty years before. The changes from a degrading to a conserved landscape are clearly visible in matching photographs taken in 1937 and 1990.

DISTRICT FEATURES

Machakos District contains central hills rising to 1500 m asl, and dropping to less than 1000 m in the surrounding plains. Rainfall reaches over 1000 mm per

Photograph 7.1(b). The same area in 1990, with fruit trees on terraces where maize has been harvested in the foreground, terracing on the slopes in the background, and a woodlot on the right. (Photo: Michael Mortimore)

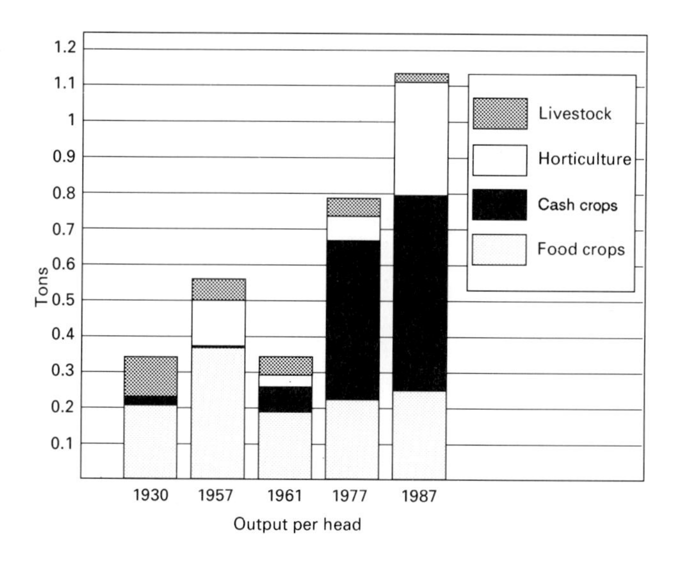

Figure 7.1. Agricultural output per head, 1930–87, converted to maize at constant 1957 prices

year on the hilltops but falls to less than 500 mm on the plains. It is distributed in two short wet seasons, the long rains (March–May) and the short (October–December). Neither may be sufficient for a crop of maize, for droughts are frequent, unpredictable, and often occur in runs of two or more consecutive seasons. Except for small areas of relatively well-watered uplands (called 'high potential' areas in Kenya), the district is semi-arid. The large-scale migration of farming households into 'low potential' zones in the 1960s and 1970s gave rise to official fears, both of ecological damage and of increased famine risk. Neither fear has proved justified.

LAND-USE CHANGE

During the century of British occupation of the Akamba country, land has passed from abundance to scarcity. This has led to changes in land use. Figure 7.2 shows some of these changes between 1948 and 1978 in three representative areas. The first area (Kangundo/Matungulu/Mbiuni) contains a high proportion of land in agro-ecological zone 3 (high-potential land). It is the most densely populated and longest settled. The second area (Masii) falls entirely in zone 4 (semi-arid upland). Its population density is intermediate and it was settled by households moving out from the hills, largely before 1940. The third (Makueni) represents the drier, warmer and lower zone 4/5 areas settled after 1945.

Figures based on air photo interpretation show the extension of the arable area from 35% to 89% in area 1, from 22% to 50% in area 2, and from 2% to 30% in area 3. The private appropriation of formerly common land (*weu*) for cultivation took place alongside the abandonment of shifting cultivation in favour of permanent, enclosed fields: a change that was encouraged by the government. It meant a reduction in the grazing areas, the increasing scarcity of which led to their appropriation also.

Figure 7.2 also shows the increase of terracing on arable land. Compulsory terracing schemes were introduced in the 1940s, but they were unpopular because the benefits were unclear. Many terraces fell into disrepair around the time of independence and the figures show that in 1961 the proportions terraced in areas 1 and 2 were essentially the same as those of 1948. Later in the 1960s, however, terracing was renewed voluntarily and farmers began to construct terraces within a few years after opening new land, even in the newly settled areas (like area 3), where average slopes are much less steep. Unterraced arable land had been virtually eliminated in all three areas by 1978, though elsewhere in agro-ecological zone 5 it might still be found. It is significant that most progress was made between 1961 and 1978, when the arable area was itself growing rapidly. Terrace construction continued in the 1980s, and was promoted by the Machakos Integrated Development Programme. By 1990 erosion was considered to be under control on arable land. The problem was, however, only beginning to be solved on grazing land.

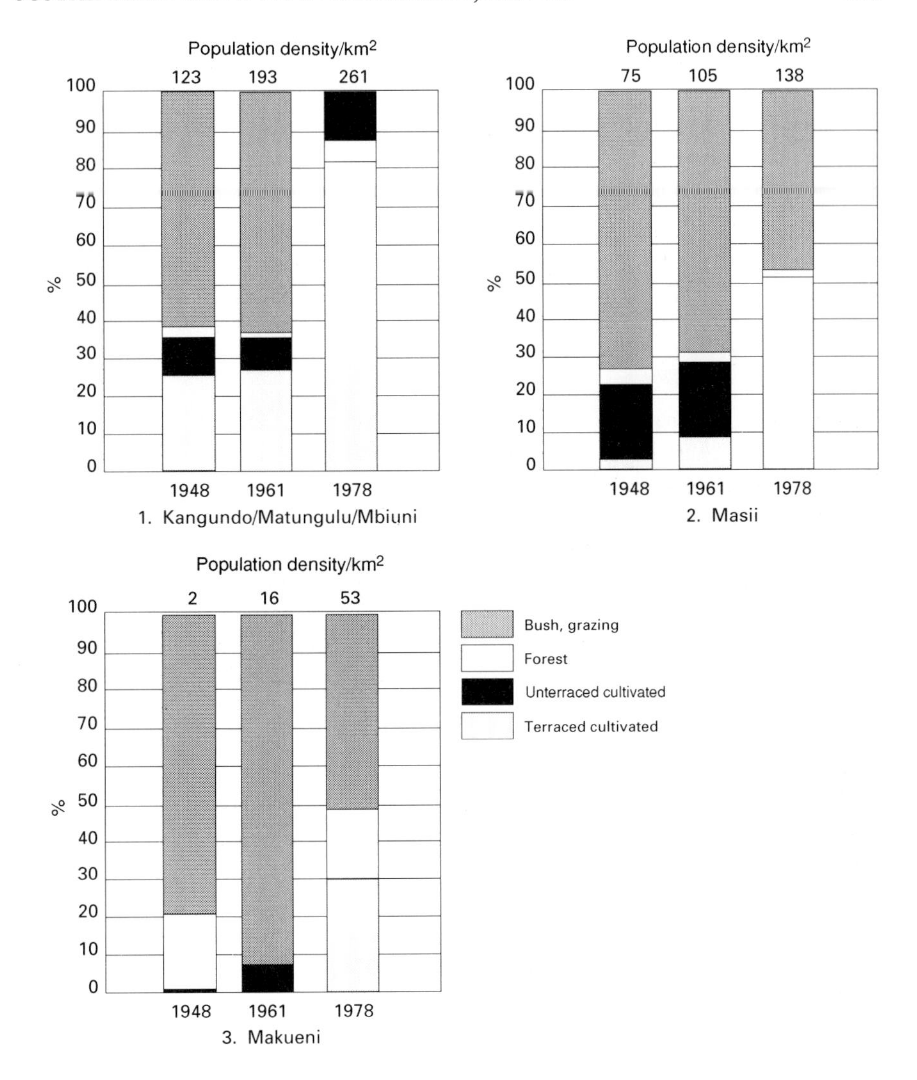

Figure 7.2. Land use in three areas, 1948, 1961, and 1978, from air photo interpretation

MAKING THE MOST OF WATER

Terracing, originally introduced as a soil conservation measure, also increases soil moisture (Table 7.1). In the late 1940s, shallow narrow-based terraces, with the soil thrown downslope (*fanya-chini*) were recommended for African smallholders. They were believed to use labour more efficiently, at a time when labour and tools were scarce, than bench terraces made by throwing the soil upslope (*fanya juu*). However, farmers began to switch over to bench terraces

Photograph 7.2(a). The Kalama Hills in the background are covered with erosion scars, and a great deal of exposed rock or soil is visible in the stream beds (1937). (Photo: Kenya National Archives)

in the 1950s. In 1961, the area conserved by narrow-based terraces had fallen to 19 000 ha (from 35 000 in 1957), partly from poor maintenance and partly due to conversion to bench terraces. By 1990, less than 5% of all terraces were narrow-based (Gichuki, 1991).

Bench terraces were desired initially for vegetables, which became profitable in the early 1950s. When African farmers were permitted to grow coffee (from 1954) they were subjected to stringent rules which included the construction of bench terraces. Now they are used for all crops. Bench terraces waste no land, as the bank can be used for fodder crops and the ditch can be planted with fruit trees. They are better at water conservation because the ditch at the rear provides water to the terraces, not, as on narrow-based terraces, to the bank. Maintenance costs are lower. The population growth and associated diminution in holding size increased the incentive to invest in the land.

A major reason given by farmers for constructing terraces was observed yield improvements. Labour, which used to be organized through government compulsion, was later provided through community self-help groups (*mwethya*), which could work on private farms in turn. It is now often hired.

Figure 7.2(b). In 1990 a great deal of the arable land has been terraced, and most of the grazing area, now privately managed, is in no worse condition than in 1937, despite the greater scarcity of this resource. (Photo: Michael Mortimore)

Table 7.1. Maximizing the efficiency of water use

Requirement	Measures
Water management	
Controlling run-off, water storage, optimal water allocation	• Water harvesting (e.g. terraces, cut-offs, pits for trees) • Supplemental irrigation • Re-use of water
Soil management	
Improving soil moisture and nutrient-holding capacity	• Controlling capping • Improving structure • Conservation tillage • Fertilization
Crop management	
Optimizing crop use of water	• Selecting drought-tolerant, drought-escaping or high-value crop varieties • Adapting cropping system • Optimizing plant populations • Weed control • Crop diversification

Terrace building calls for substantial labour mobilization for short periods and may therefore be beyond the resources of the household, unassisted. Much, if not most, labour is provided by women, as men are more involved in off-farm work.

Water scarcity, whether for domestic, livestock or crop use, results from the low and variable rainfall, high run-off losses, momentary stream flow, low groundwater yield, low moisture-holding capacity in soils and high rates of evaporation. Most critical is water availability for arable farming. Various strategies are available for maximizing the efficiency of water use. Smallholders in Machakos learn from their neighbours, extension officers and their own experiments. For example, on the 2 ha farm of Mr Musyoki, in agro-ecological zone 4, the following measures may be observed: cut-off drain with bananas planted in pits, bench terraces, diversion of roadside runoff for crop use, conservation tillage, mulching, weed control and manuring, mixed cropping with fruit trees, beans and maize, live fencing used as a windbreak, supplemental irrigation from pond storage for vegetables and the use of grafting to diversify and improve fruit yields (Figure 7.3). Mr Musyoki used to be a waiter in Nairobi, and he started farming on degraded land in 1970.

RECYCLING SOIL NUTRIENTS

The soils of Machakos are naturally deficient in phosphorus. Repeated cultivation without fertilization reduces nitrogen, carbon and exchangeable cations to low levels (Table 7.2). Even long fallowing under grazing fails to restore soil nutrients to the levels found on uncultivated land. Fertility management is therefore critically important for the sustainability of arable farming.

There are four main soil-improvement options: inorganic fertilizers, boma (farmyard) manure, alternative organic sources (compost, mulches, green ma-

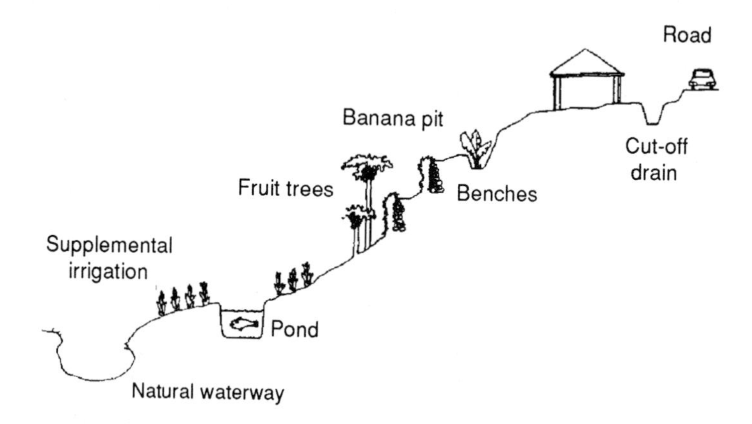

Figure 7.3. Mr Musyoki's farm demonstrates ways of conserving water and increasing crop output which are now common in Machakos

Table 7.2. Chemical properties of soils under three management regimes

Property	Long-term management regime		
	Uncultivated	Grazing	Cultivated
Soil pH (water)	5.5	5.4	5.0
Potassium (meq)	0.56	0.4	0.29
Calcium (meq)	8.7	2.4	1.1
Magnesium (meq)	3.4	1.4	0.9
Phosphorus (ppm)	23.0	14.0	13.0
Nitrogen (%)	0.35	0.18	0.11
Carbon (%)	2.49	1.25	0.74

Note: Soils with known management histories in Kilungu location were sampled from the top 20 cm at three sites under each management regime in each of the three areas (27 composite samples in all). Source: Mbuvi (1991).

nure), and use of nitrogen fixing legumes as intercrops, rotations or farm trees (Mortimore and Wellard, 1991). Except for the use of nitrogen fertilizers on the profitable coffee crop, manure has been the mainstay of soil-improvement efforts since the 1930s and 1940s, when it acquired a commercial value and the Department of Agriculture started promoting it. Adoption is now nearly complete, in terms of the numbers of farmers using it. However, supply constraints (number of livestock or working capital available) restrict its application on most farms to levels well below those desired for optimal nutrient supply.

Composting offers an addition to manure and by combining manure with plant materials, it can double the supply of organic material available. It was first promoted by the Department of Agriculture in the 1930s, but did not become popular then. Recently, it has been taken up again by NGOs working in the district. In a scheme run by the Kenya Institute of Organic Farming, farmers' groups combine their labour for compost making using, among other materials, the invasive weed *Lantana camara*. The system represents a further step in labour intensification.

Manure and compost improve nitrogen and phosphorus levels, soil structure and moisture retention. Leguminous crops are already important. Inorganic fertilizers usually cannot be justified economically, except for high-value crops such as coffee (138 kg/ha in 1988). Although use on food and horticultural crops increased in the 1980s, averages are still very low (1.6 kg/ha in 1988). Inorganic fertilizers have always been recognized as a second best to manure.

It is not known what quantities of manure are applied in the district. However, since the 1930s, the output of food and cash crops and horticulture has increased by a factor of 11 per hectare, while the area of arable land has grown by a factor of six (Rostom and Mortimore, 1991). The (inadequate) data indicate that crop yields did not fall, and probably rose, both for maize and for the market crops (Mbogoh, 1991). Farmers combine a research-derived drought-tolerant maize variety with local varieties.

FEEDING LIVESTOCK

Livestock data are weak and contradictory. However, the number of stock units (1 unit = 1 cow or 5 sheep/goats) seem to have risen from just over 200 000 in the 1940s, a time of frequent droughts and high mortality, to at least 334 000 and possibly as many as 588 000 in the 1980s (Ackello-Ogutu, 1991). Allowing for the change in area, overall densities fell from 2.4 ha/unit in the 1940s to 4.4 in the 1980s (if the lower figure is accepted) or remained virtually the same at 2.5 (if the higher is accepted). In high population density areas the need for more productive feeding systems was unavoidable as, owing to subdivision on inheritance, the average size of Akamba holdings is constantly diminishing. The transformation of livestock feeding systems, which is yet incomplete, consists of three linked steps (Mortimore and Wellard, 1991).

First, the private appropriation of almost all common land forced a matching of stock holdings with carrying capacity at the level of the individual farm. Otherwise degradation threatened the long-term viability of the holding, a critical consideration in planning for one's heirs. In agro-ecological zone 5 no evidence was found of grazing-induced degradation of the natural vegetation. Rather, the proscription of fire as a management tool has allowed the increase of woody species on rangeland where grazing has been introduced mainly during the last three decades. On private land with a long history of intensive grazing in zone 4, farmers employ destocking, grass planting, control of unwanted species and tree protection to manage the ecosystem sustainably. The stabilization of formerly mobile grazing circuits within the bounds of private holdings increased the use of crop residues as fodder. Maize stalks and bean haulms are used universally for this purpose.

Second, fodder grasses and trees are now grown on terrace banks and farmlands respectively. Fodder grasses were promoted by the government as long ago as the 1930s. In the early 1950s it was shown in a government experiment, at the 'Makaveti square mile', that such grasses grew well even on degraded land, and that they could support much higher cattle densities than had been thought possible, namely up to 0.4–1.6 ha per livestock unit (Ackello-Ogutu, 1991). Today, the grasses *Pennisetum purpureum* (Napier grass) and *Panicum makarikariensis* play an important role in integrating the crop and livestock sectors and intensifying the farming system.

Third, stall feeding of livestock is practised now by 59% of cattle, 53% of sheep and 50% of goat owners, respectively, for part or all of the year. This method permits more efficient use of residues and fodder crops and more systematic use of manure. It disposes of the problem of where to put livestock during the growing season, on crowded arable holdings, and protects terraces from damage by grazing animals.

Ecology and population density influenced the adoption of both fodder production and stall feeding. In Kangundo (area 1 in Figure 7.2, and in agro-ecological zone 3), almost all farmers are now reported to plant fodder, and stall feeding is the most common feeding system. In Masii (area 2, and in zone 4), only a few plant fodder or practise stall feeding. In Makueni (area 3 and in zone 4/5), while some plant Napier or Rhodes grass, stall feeding was tried earlier but abandoned. In Ngwata, in zone 5, no farmers do either (Mortimore and Wellard, 1991).

FINDING CAPITAL

Even when land was abundant, capital was necessary for its development. As land scarcity increased, capital supply became critical in the adoption of technologies of intensification and conservation: the construction of terraces and other soil and water conservation structures and enclosures, the purchase of fertilizers, the acquisition of ploughs or fixed farm capital, entry to coffee production, the purchase of improved livestock and access to markets.

The experience of Machakos supports the Boserup hypothesis that increasing population density leads to intensification through changing labour-to-land ratios. However, labour intensification cannot be isolated analytically from capitalization. Labour for land improvement and tree planting was hired, as well as supplied from family and communal sources. Not only does capital substitute for labour, but in many technologies it supplements it. For example, terrace construction requires tools, effective manuring and composting require ox-carts and water drums, grade livestock require frequent dipping, coffee and vegetable crops need purchased inputs.

The Community Development Department (under the leadership of John Malinda) adapted the traditional *mwethya* work party, used to create new farms or houses, to the needs of soil conservation in the 1950s. The *mwethya* groups appointed their own leaders and worked on their members' farms in turn. Leaders and musicians were often women, since so many men were away working. Nowadays they are used to raise both capital and labour, for community as well as private projects—schools, bridges, businesses as well as terraces—while NGOs are using them to identify, plan and carry out projects. The *mwethya* groups, therefore, have made a major contribution to the capitalization of the economic landscape (Mbula Bahemuka and Tiffen, 1991).

However, capital was also derived from farm profits (sales of livestock and crops) and savings from waged work or non-farm enterprises. We were unable to measure the respective contributions of labour and cash investments, but there is no doubt that the latter were substantial.

CONCLUSION

Change, intensification and conservation were driven by population growth, increasing land scarcity, investment, and new information and technologies. In livestock management, the evolution from extensive grazing in the 1930s, through increasingly labour-intensive methods, to the growing popularity of stall feeding in the 1990s illustrates the general trend towards intensification. In the process, the Akamba have come to value livestock not only as a savings reserve, which can be utilized in emergencies or to provide investment capital, but also as sources of farm energy, manure and regular income from milk and meat sales.

It is a major conclusion of this study that the market facilitated and promoted change. In the 1930s, better-off farmers invested in ploughs to produce maize for the market. Coffee generated investment income in the high-potential areas from the 1960s. Canning factories and traders were among those providing credit, incentives and information for fruit and vegetable production. However, it is questionable whether reinvestment of agricultural profits could alone provide all the capital needed to transform the farming system. To a certain extent, capital can and did take the form of direct labour investments, by family labour or groups. The external labour market was also an important source of both capital and information. Many Akamba joined the army and police, travelled widely and brought back savings and new ideas. Education was seen as giving access to employment outside the district and outside the farming sector, the value of which in years of crop failure and livestock mortality was apparent. Households used income diversification both to secure themselves against risk and to build up savings for investment in the farm (as illustrated by Mr Musyoki's farm). Other investments, of work and cash, were in community infrastructure, such as feeder roads and schools.

A typical investment cycle was to seek market participation, either for labour or for farm products, to generate income. This income was used to educate children. Some children (predominantly males but, increasingly, females also), on attaining maturity, took jobs inside or outside the district, sending part of their earnings to support the family and/or to finance investments. Productive investments increased farm incomes and the value of the land. The holding was subdivided on the owner's death among heirs into smaller but by then more fertile parts, and the cycle was repeated.

The most important implication of this development is that technological change was functionally linked with income diversification and increased market participation: the sustainability of the farming system cannot be considered in isolation from the household economy as a whole. The relatively small areas of high value crops, typified by Mr Musyoki's vegetable plot, help to generate the incomes which enable farmers to improve their houses and to purchase more goods and services, thus creating more off-farm employment locally. By 1981–2 a national survey showed that in Machakos only half the rural income was generated by the farm, the rest by other activities. In some of the more densely populated districts of Kenya, with higher average incomes than Machakos, the proportion of non-farm income is even higher.

WHAT FUTURE FOR MACHAKOS?

Although typical family size is now being reduced in Machakos, population will inevitably continue to increase, given the age structure. There is still scope for agricultural intensification in parts of the district, which could be facilitated by improving market access. Output per hectare in a given agro-ecological zone is far higher on small farms than on large ones, as the small farms use more working capital per hectare (Tiffen, 1992). At the moment, the insufficient road network is decaying, limiting marketing opportunities for milk and vegetables and reducing farm-gate prices on livestock and other products. People are energetically seeking new non-farm income sources. Craft industries, such as the manufacture of the well-known Machakos sisal basket, woven by women, or the wooden carvings made by men, are expanding. Village leaders we interviewed in a very densely populated area said that they needed rural electrification to enable them to process more agricultural products, to establish more workshops catering for local needs and to generate more employment. This is happening in the few places in Machakos that have good roads and an electricity supply.

Machakos will change again in the future. Given policy support, it will become more urbanized and industrial. However, there is no reason to think that this will undermine the sustainability of its agricultural base.

NOTES

1. This article first appeared in the *ILEIA Newsletter,* 9, No. 4, 1993, and the authors wish to thank ILEIA for permission to reprint it here.

REFERENCES

The material on which this chapter is based is reported and referenced in: Tiffen, M., Mortimore, M. and Gichuki, F. (1994) *More People, Less Erosion: Environmental Recovery in Kenya.* Chichester, John Wiley.

ODI working papers (WP) contain more details on some aspects. Those quoted are:

ODI WP 53 (1991) Environmental Profile. Mutiso, S.K., Mortimore, M. and Tiffen, M., 'Rainfall'; Thomas, D.B., 'Soil erosion'; Mbuvi, J.P., 'Soil fertility'; Farah, K.O., 'Natural vegetation'.

ODI WP 55 (1991) Production Profile. Ackello-Ogutu, C., 'Livestock production'; Mbogoh, S.G., 'Crop production'.

ODI WP 56 (1991) Conservation profile. Gichuki, F.N.

ODI WP 57 (1991) Profile of technological change. Mortimore, M. and Wellard, K.

ODI WP 58 (1991) Land use profile. Rostom, R.S. and Mortimore, M.

ODI WP 59 (1992) Farming and income systems. Tiffen, M.

ODI WP 62 (1992) Institutional profile. Mbula Bahemuka, J. and Tiffen, M.

ODI WP 63 (1992) Tree management. Mortimore, M.

8 Natural Resource Management in Pastoral Africa

ROY BEHNKE
Overseas Development Institute, London, UK

Current research on rangeland ecology suggests that we have less to fear from pastoral land stewardship than was previously thought. On the one hand, the natural environments exploited by pastoralists are generally robust and resilient. On the other, pastoral techniques of land management are not as dysfunctional as was once widely assumed. While regulation of pastoral activity may be necessary in specific circumstances, there no longer exists a broad scientific mandate to control or modify almost every aspect of pastoral land use in order to preserve the environment.

This brief chapter tries to do two things. First, it attempts to justify in some small measure the assertions made in the above paragraph. Then it examines the fundamental adjustments in natural resource management policy which would be required if these assertions were substantially true.

THE RESILIENCE OF AFRICA'S RANGELANDS

In the highly 'pulsed' environments which concern us here, there are marked seasonal and inter-annual fluctuations between wet and dry periods. During the wet seasons/periods when the vegetation is growing it is vulnerable to damage by herbivores, but it also tends to be in considerable surplus to their feed requirements. Conversely, when grazing pressure is high—i.e. in dry seasons or during droughts when demand for feed far outstrips supply—the plants tend to be relatively impervious to abuse, the living parts having retreated behind thorns, inside woody structures, below ground or stored in seeds.

In sum, in these very harsh and erratic climates the plants are relatively well adapted to high grazing pressure during dry periods of forage scarcity. And dry periods are frequent enough and severe enough so that it is difficult to maintain domestic herbivore populations of sufficient size to challenge the vegetation when it is growing and vulnerable. These are ecological systems with a considerable capacity to absorb perturbation.

Social Aspects of Sustainable Dryland Management. Edited by Daniel Stiles

If this is true, policy concerns about overstocking and desertification may be justified in particular instances, but these phenomena are not general enough to provide a framework for the formulation of natural resource management policy as a whole. A more worthy focus of policy attention are the peculiar problems of pastoralists who have to make a living in dry and drought-prone environments. These problems are examined in the next section, which compares resource management and exploitation in erratic or 'non-equilibrium' versus stable or 'equilibrium' rangeland settings.

MANAGEMENT IN EQUILIBRIUM AND NON-EQUILIBRIUM GRAZING SYSTEMS

Equilibrium grazing systems are characterized by relatively high levels of climatic stability resulting in constant levels of primary production. Provided with a reliable feed supply, livestock populations in these settings can expand to the point where they have a considerable impact on the vegetation, in terms of both its species composition and density.

Conventional range management makes sense in this kind of grazing system. Livestock make spaces either by eating or stepping on plants; plant succession tends to fill these spaces back in. Range management seeks to balance these space-creating and filling processes by adjusting the destructive power, and hence the number, of animals to match the recuperative power of the plants, thereby sustaining a stable and profitable equilibrium. In this model 'fine tuning' for optimal output is possible because livestock renew themselves—reproduce, grow and produce meat and milk—at a rate determined by the availability of their feed, which is an inverse function of their density. Changes in stocking density therefore yield predictable alterations in output and revenue.

Implicit in this model is the assumption that managers are able to exercise considerable control over future biological states of the grazing system. They are able to do this because the variables which drive the system are biological components of the system itself and are susceptible to human control at relatively low cost. This is an appealing message for commercial ranchers who occupy marginal land which is incapable of generating the revenue to sustain ambitious capital investments or the intensive employment of labour. However modest their profit margins, here is one 'management' operation which every rancher should be able to afford—the harvesting for sale of animals from his herd. Conventional range management invites the manager to transform this simple profit-taking operation into a sophisticated management tool by setting a culling rate which will leave behind an optimal number of animals.

But what happens when it does not rain or, more precisely, what happens in areas where rainfall is persistently erratic both in timing and spatial distribution, the so-called non-equilibrium environments of arid and semi-arid Africa? In a grazing system subjected to extremely variable rainfall, fluctuations in

Photograph 8.1. Dryland plants are adapted to high grazing pressures and perturbations such as trampling. (Photo: Daniel Stiles)

rainfall may have a much stronger effect on the vegetation than animal numbers. Productivity levels will therefore be determined by abiotic perturbations which the manager is unlikely to be able to predict and will, almost certainly, be unable to control. In these environments it is no longer appropriate to conceive of management as the manipulation of the biological system to achieve maximum output or revenue. Such manipulation is not possible because the critical variables which drive the system are not susceptible to control; moreover, extreme fluctuations in output and revenue—not the imprecision with which these values are maximized—are the primary causes of distress for dependent human populations.

ECOLOGY AND RESOURCE TENURE

In an equilibrium grazing system it can be quite useful for a single decision-maker to control all the animals in a delimited area. This level of control permits managers to make precise adjustments in livestock numbers and, in most situations, equates an owner's self-interest with the maintenance of an appropriate long-term stocking rate. Restrictions on herd movement present no insuperable difficulties as long as the supply of feed produced within an area is relatively constant, predictable and susceptible to manipulation by adjusting the stocking rate.

The situation changes, however, as the climate becomes more erratic and the manager discovers that he cannot manipulate his forage supplies, but is

Photograph 8.2. When one area has been grazed down, livestock are moved to areas of pasture surplus. This bare ground will regenerate with the rains and be able once again to feed livestock. It is debatable whether repeated episodes of such overgrazing lead to long-lasting degradation. (Photo: Daniel Stiles)

manipulated by them. The basic problem may remain the same—that of matching feed supply and demand—but it manifests itself very differently. In the equilibrium setting, the challenge was to maintain a long-term balance by making almost continuous minor adjustments to livestock numbers. In the non-equilibrium setting, the problem is to react quickly or to 'track' unpredictable, dramatic but short-term fluctuations in feed supply. One effective tracking strategy is to move animals around, avoiding areas where forage is insufficient and mopping up surpluses in areas where it is abundant.

Pastoralists living in a very unstable environment will, therefore, derive little advantage from having unquestioned control over a sporadically productive stretch of territory; more advantageous would be the option to use alternative areas, should these areas be capable of sustaining their livestock when their home territory is not. But options on another person's property will, in all probability, mean that they have a reciprocal lien on one's own property. This *quid pro quo* exchange of use rights is the basis for the non-exclusive tenure and land use regimes characteristic of pastoral Africa.

The inversion of our expectations regarding responsible resource management could not, in some respects, be more complete than it is in pastoral Africa. Whereas freehold ranchers may be forced to prudently adjust livestock numbers to their land base, pastoralists tend to reverse this equation. Faced with large and unpredictable fluctuations in resource productivity, they seek to lay claim to the natural resources which are needed to sustain their stock. They fiddle their tenure system rather than their herd size, thereby escaping the 'salutary' discipline of freehold tenure, that certain knowledge that they must live and operate within a rigidly delimited and finite land base.

Economic theorists, planners and professional range managers all find this rather disturbing. The setting of rigid ranch boundaries and the unequivocal identification of ranch members were both a persistent concern for designers and a persistent problem for implementors of the older generation of group ranches (Perrier and Craig, 1983; Perrier, 1990; Devitt, 1982). Things have not changed much in the intervening years; current research on common property resource management endorses precisely these concerns. Both theoretical (Runge, 1986) and empirical studies (Ostrom, 1990) indicate that the unambiguous identification of a delimited group with a bounded resource is a critical ingredient in the success of collective resource management. What is required for effective management, this work suggests, is not necessarily private tenure or an individual manager but rather exclusive tenure (even if rights are exclusively owned by a group) and a single decision-making body (which may be corporate).

These conditions for successful group management cannot be met on any widespread basis in dry Africa. 'Fuzzy' or indeterminate group and territorial boundaries are not an inadvertent feature of pastoral tenure, or simply a residue of a time when population densities were low and the precise definition of property and property holding groups was unnecessary. In an unpredictable environment, certain critical ambiguities as to who owns what and can go where provide a degree of fluidity which suits everyone's requirements.

SOME PRACTICAL SUGGESTIONS

So, you may say, we now have a better understanding of how pastoral resource management systems work. But are we any closer to understanding how we—as external agents of development—should intervene in these systems? There follow a few points for debate:

(1) The per hectare productivity of Africa's arid and semi-arid rangelands is low; so must be the cost of their management, if management is to be sustainable in the sense of generating improvements in production which equal or exceed the costs of management. Modest expectations are in order.

(2) In a difficult and complex environment the energetic administrator is constantly tempted to overstep the economic limits of appropriate administrative endeavour by managing at an unaffordable level of intrusiveness, precision and intensity. For this reason it is quite unlikely that we will

discover suitable models of pastoral administration by examining livestock and range development projects funded by donor agencies. Indeed, these projects may have fundamentally distorted our expectations of what pastoral resource management can achieve on a continuous basis, and the resources which are likely to be routinely available.

A more realistic framework for sustainable management effort is provided by the typical district-level bureaucracy in dry Africa—the colonial district commissioner or his independent counterpart, and their local subordinates—beset by a plethora of problems which demand attention, constrained by inadequate staff, limited funds, poor communications and transport, and encumbered by superiors at headquarters who are only sporadically concerned with local affairs. Set resource management objectives which are consonant with the capacities of these officers and we will, in all probability, be working within realistic limits.

(3) Not much administrative intervention in resource management is feasible within the constraints laid down here. If the landscape is to be managed at all, then individual pastoralists and pastoral communities will have to do it, as they have in the past. If these local users are to manage responsibly, then they must be given legal rights to control the resources they are managing. This requires governments to relinquish *de jure* control over rangeland resources; control which in any case was effectively beyond their grasp. In relinquishing this control administrators and donors can be reassured that, as was noted at the outset, we now have less to fear from pastoral land stewardship than was once assumed.

(4) Since regulatory aspirations are unrealistic and unnecessary, administrative emphasis should shift from regulating resource use to allocating and upholding access rights. Administrators/managers can usefully arbitrate between the conflicting interests of opposed groups and individuals. The need for such mediation will be fairly constant in an erratic environment in which resource productivity, people and animals are continually rearranged in space, and land-use issues cannot be sorted out once and for all. Indeed, the arbitration of chronic conflict over scarce resources may be the central natural resource management function of local government officials in a non-equilibrium setting.

Given the limited means at their disposal, government officials probably can only afford to make management decisions at the margins, when the ownership and use of a particular resource hangs in the balance between contesting parties, and the administrator can expect to exercise a degree of influence which is disproportionate to his real power. It is at this modest level—with the ability to influence but not to dictate landholding patterns and management practices—that the typical district official must learn to operate.

(5) Continuous official involvement in arbitrating land tenure disputes effectively sidesteps the problem of how to transform customary land rights

into formal law. Pastoral tenure regimes exhibit a level of complexity and internal variability which is virtually impossible to simplify into legal formulae. Even if it was initially accurate, such codification would soon be out of date, as customary usages change in response to demographic pressure, new economic interests or shifting political alignments. The essence of customary tenure is its fluidity, and this facet of customary usage cannot be captured in law by simply writing down existing oral agreements.

One solution to this impasse is for government authorities to promulgate and subsequently enforce procedural rather than substantive land law (Vedeld, 1993). Rather than legislatively dictating the content of property rights, procedural law would specify the framework within which interested parties could legitimately put forward alternative claims to resources, what administrative/jural institutions should process claims, the criteria for choosing between opposing claims and enforcement procedures.

Donor support for the development of procedural law would imply the renunciation of any attempt to dictate the kind of tenure system—individual or communal—which was appropriate in a particular locality.

(6) And where, literally, should the authorities begin their management activities? With 'key' resources—the natural resource bottlenecks, the scarce natural production factors which limit livestock output by controlling stock numbers or productivity during the most difficult periods in the production cycle. Which natural resources were key or limiting within particular production systems could be determined by applied research, or by simply observing which resources occasioned chronic conflict and dispute.

Focusing on certain resources means ignoring others. In focal point management the intensity of management activity, investment levels and the precision with which access rights are specified would be intentionally adjusted to reflect the value of the resource being managed. In contrast to the freehold model of individual or group ranch tenure, focal point management would concentrate development efforts on a particular category of resources, rather than the delineation and management of bounded territories containing a variety of different resources. In essence, focal point management would concentrate management attention on these resources which lay at the heart of the production system, and devote much less effort to the clarification of rights to resources which were very abundant, of low or erratic productivity, or geographically extensive and difficult to police.

The concentrated deployment of management effort offers several advantages. Most importantly, it would ensure that government's scarce financial resources were expended where they were most needed, on the management of resources which were in high demand, were likely to be a source of conflict between different groups of users, and—if left unmanaged—could precipitate disputes which threaten public security.

Producers who controlled key natural resources would possess a resource base which permitted them to exploit more peripheral resources and exercise *de facto* control over these resources. Thus, through the allocation of critical nodes on the landscape, administrators could expect to exercise some indirect influence over levels of resource exploitation throughout an area or region. Finally, by de-emphasizing the need for strict boundary maintenance, focal point management would permit the continuation of customary tenure arrangements which encourage the shared use of resources which are not in high demand.

NOTE

This paper is an abbreviated version of an article which appeared in *Development Policy Review*, 1994, **12**, 5–27.

REFERENCES

Devitt, P. (1982) The management of communal grazing in Botswana. *Pastoral Development Network Paper* 14d. London, Overseas Development Institute.

Ostrom, E. (1990) *Governing the Commons: The Evolution of Institutions for Collective Action*. Cambridge, Cambridge University Press.

Perrier, G.K. (1990) The contextual nature of range management. *Pastoral Development Network Paper* 30c. London, Overseas Development Institute.

Perrier, G.K. and Craig, P.S. (1983) The effects of the 'development approach' on long term establishment of a grazing reserve in Northern Nigeria. *Pastoral Development Network Paper* 15c. London, Overseas Development Institute.

Runge, C.F. (1986) Common property and collective action in economic development. In: *Proceedings of the Conference on Common Property Resource Management*. Washington, DC, National Academy Press and the National Research Council.

Vedeld, T. (1993) Oral contribution to *New Directions in African Range Management and Policy Workshop*. Woburn, Commonwealth Secretariat, Overseas Development Institute and the International Institute for Environment and Development.

9 Indigenous Peoples, Resource Management, and Traditional Tenure Systems in African Dryland Environments

ROBERT K. HITCHCOCK
University of Nebraska, Lincoln, USA

INTRODUCTION

A recent trend in the analysis of African systems of land and natural resource management and user rights to resources is the consideration of community-based systems of resource management (Lawry, 1990; Associates in Rural Development, 1992). Land and natural resource use and management in Africa includes the growing of crops, raising livestock on open ranges and fenced ranches, mining of economically important minerals, hunting of large mammals, gathering of wild plants for subsistence, income generation, manufacturing, and medical purposes, rural and urban industries based on natural resources, conservation of endangered and rare species, and protection of significant habitats (Office of Technology Assessment, 1984; Anderson and Grove, 1987; Hitchcock, 1993). With increased population and expanded infrastructure and development activities in both rural and urban areas, there has been an intensification of natural resource exploitation and an increase in conflicts among various forms of land use (Timberlake, 1988).

In the past several years, suggestions have been made by researchers, government agencies, and development organizations that natural resources could provide a sustainable source of employment opportunities and rural income generation in Africa (Harrison, 1987; Stiles, 1988; USAID, 1989a,b; Pradervand, 1989). Projects incorporating community-based resource management were planned and are in the process of being implemented; these projects range from game cropping schemes to tourism and from rural industries to small-scale livestock loan schemes (for a list of some of these projects, see Appendix 9.1). Analysis of these projects indicates that sustainable development can only be achieved if careful attention is paid to local people's participation in

Social Aspects of Sustainable Dryland Management. Edited by Daniel Stiles

decision making, strengthening of resource management institutions, a multi faceted approach to economic promotion, and environmental conservation and education efforts (Harrison, 1987; Durning, 1989).

One approach that has been suggested recently is that African countries, with the help of local communities, donors and non-governmental organizations, should implement projects that will diversify economies and expand incomes. It has been recommended that wildlife and wild plant products be managed in sustainable ways so that they can play more significant roles in expanding rural economies and enhancing the standards of living of local people (Stiles, 1994; Associates in Rural Development, 1992). Given the importance that many countries in Africa place on the environment and rural development, it was appropriate that national policies on development, conservation and sustainable resource utilization should be developed and put into effect.

LISTENING TO THE PEOPLE: COMMUNITY EMPOWERMENT AND PARTICIPATION

The issues of community empowerment and local participation in rural development projects are receiving increasing attention from researchers and development organizations (Cernea, 1985; Midgley, 1986; Associates in Rural Development, 1992). Participatory development has become a catchphrase for the kind of approach that many agencies and policy analysts are advocating. Various means of bringing about local participation have been advocated, including provision of training and education (investment in human resources), and ensuring that local people have control over their own land and natural resources.

The concept of participation is one that is not easy to define. It can mean the right to make decisions about development action. Participation can also mean the process whereby local communities take part in defining their own needs and coming up with solutions to meet those needs. In addition, participation can refer to situations in which local communities and individuals share in the benefits from development projects and are fully involved in generating those benefits. As Chambers (1983: 140) notes, 'Rural development can be redefined to include enabling poor rural women and men to demand and control more of the benefits of development'.

The community development approach advocated by various agencies in the 1950s and 1960s used a definition of community development which is pertinent:

> Community development is a process of social action in which the people of a community organise themselves for planning and action; define their common and individual needs and problems; make group and individual plans to meet their needs and problems; execute the plans with a maximum of reliance upon community resources; supplement these resources when necessary with services

Photograph 9.1. Education is an important prerequisite to bringing about effective local participation, as here with the Basarwa of eastern Botswana. (Photo: Daniel Stiles)

> and materials from Government and non-Government agencies outside the community (International Cooperation Administration, 1955: 1).

Essentially, participation and empowerment are similar to this kind of approach, especially when emphasis is placed on self-help, self-determination, and local control of resources and decision making. In order to overcome development constraints, it is important to promote strategies of community consultation, mobilization, and organization. This process must also include ensuring that those communities have the ability to maintain control over their human and natural resources.

It is important to remember that the degree of willingness of individuals to take part in development action and to take responsibility for decision making often various tremendously, not only within specific areas but frequently within the same family. In order to determine the various goals and objectives of local people, concerted efforts have to be made to collect information and seek feedback at the local level. What this means is that an investigatory programme must be built into all development projects. It also means that continuous monitoring and consultation has to be done during the course of project identification, design, implementation and evaluation. If it is found that local people do not agree with the ways in which the projects are designed or being put into practice, then changes must be made or new approaches must be taken.

A number of rural development and natural resource management projects have taken an approach in which communities contribute resources of their own to the development efforts, including cash or payment in kind (e.g. labour). They have avoided handouts, which also encouraged the strengthening of local institutions so that they could make decisions on their own. In some cases, this has meant that changes had to be made in national legislation so that the institutions could manage their own finances. Promotion of literacy and numeracy among local people has been a key strategy as well.

Some governments and non-governmental organizations have developed what they term a participative extension approach to rural development (e.g. the Groups Development Programme of the Ministry of Agriculture in Botswana and the Boscosa Project of Conservation Foundation and WWF on the Osa Peninsula in Costa Rica). This kind of approach places emphasis on community involvement in all aspects of project design and execution. In some instances, this approach results in the formation of local organizations (e.g. farmers' assocations or women's multi-purpose development groups). It also contributes to situations in which efforts are made to provide local communities with rights over land and other resources. Agroforestry projects, for example, are being done more and more at the household level, and tenure rights are being defined in such a way that individuals and groups have *de jure* rights, rather than simply *de facto* control.

One strategy to promote participation has been to appoint local people as 'change agents' or 'community extension workers'. By having these people at the grassroots level, it is possible for trust to be built up and for detailed knowledge about local situations to be drawn up for assisting communities. It is crucial, however, that these individuals are not seen as trying to direct the process of development; rather, they must be viewed more as facilitators and serve as advisers or information disseminators. These individuals sometimes serve as a link between community organizations and outside agencies. In this capacity, they can provide a kind of communication function.

Another strategy of empowerment and promotion of participation is institution building or institution strengthening. Most, if not all, local communities have informal associations of people who have common interests and/or who co-operate on various tasks. Peasant communities have co-operative labour units and informal working arrangements among members who share in agricultural labour or other activities (e.g. construction of storage facilities). Pastoral populations have social arrangements whereby livestock is cared for and managed for individual stock owners who allow the caretakers to use the products and energy of those animals. Some agricultural societies have groups which manage water (e.g. in Oaxaca in Mexico and in Iran). There are also voluntary associations such as producer and marketing groups in many places such as West Africa and South Asia. These institutions can be used as the basis for promoting development at the community level.

It has sometimes been said that local elites or extant authority structures often get in the way of participatory development. One way of getting around this problem is to involve the elites and local authorities in the development process. This was done among chiefs and their councillors in Swaziland in the traditional sector development programme, part of the USAID and Government of Swaziland manpower development project (SWAMDP), and among traditional healers in Nigeria, and it proved to be reasonably effective. Consulting local leaders at all phases of project formulation and implementation enables communities and development organizations to obtain information, and it helps to ensure that the leaders are fully aware and supportive of programme activities.

The building of capacity for local decision making can be done in a number of ways. It can be brought about through the holding of workshops or community discussion sessions in which ideas about democratic processes of public policy formation are addressed. It can be promoted through training of various kinds (e.g. in how to form committees, draw up constitutions and run meetings). It can also be facilitated through problem-solving exercises, case studies and role plays about situations in which communities find themselves. These kinds of strategies have been very effective in Central and South American rural communities, among women's groups in Africa, and among farmers associations in Asia (Cernea, 1985; Economic Development Institute, 1987).

MODELS OF COMMUNITY EMPOWERMENT AND PARTICIPATORY DEVELOPMENT

There are relatively few examples of truly participatory development and community empowerment programmes and projects in which local people have been fully involved in processes of change. One reason for this situation is that often development projects have short life spans, whereas institutional development and community empowerment requires long periods of time and a great deal of patience. Another reason is that often the development or conservation programmes being advocated do not lay the groundwork necessary to ensure that the local people have a stake in the projects. They also do not gain the support of the government so that legislation is changed to make possible local control over resources.

A third reason that participatory approaches to development are overlooked is that easily definable project outputs such as infrastructure construction or agricultural yield increases are given preference over less precisely quantifiable indicators such as institutional strength and resource management capacity. Often, greater emphasis, funding, and technical support are given to outside agencies (e.g. contracting groups, non-governmental organizations) than to community-based organizations (CBOs). If local communities are to be empowered and participatory development actually carried out, then there will have to be a significant change in the way that development agencies, donors and voluntary organizations deal with local people and their concerns.

Coming up with successful models of community empowerment is by no means easy, particularly since there are relatively few detailed, analytical studies that exist. One hears, for example, of the strategies employed by the Aga Khan Foundation in promoting village organizations in Pakistan and the village corporate models advocated by Norman Reynolds in southern Africa, but finding data on these models is difficult. Much of the work done on these issues has either been carried out at a theoretical rather than a practical level or it exists in the so-called grey literature of development agencies and NGOs. It is necessary, therefore, to draw upon examples for which a fair amount of information exists.

Some of the more forward-thinking approaches to participatory development in the world are being done in Africa. As a continent, Africa is diverse, with fifty countries ranging in size from 300 km^2 (the Maldives) to 2 205 810 km^2 (Zaire) (see Table 9.1). There are over 750 million people in Africa, and the population is increasing rapidly.

The vast majority of Sub-Saharan Africans are farmers, most of whom reside in rural areas. Many people make their living through a combination of agriculture, domestic animal keeping and wage labour. Approximately 24 million Africans are herders who raise livestock both for subsistence purposes and for sale. The urban population of Africa is increasing rapidly, so much so that cities such as Lagos and Nairobi have experienced serious shortages of housing, employment and social services.

The states of Africa today are largely multi-ethnic entities that are controlled by indigenous elites who vary greatly in size and cultural characteristics. Some countries, such as Swaziland, are occupied almost entirely by a single ethnic group, but these groups are usually subdivided along lines of kinship and social affiliation. Nigeria, on the other hand, contains as many as 160 different groups. The picture is complicated by the fact that the various African societies speak as many as 2000 different languages and have an array of religious beliefs.

Many African governments are reluctant to acknowledge the existence of distinct indigenous groups within their boundaries. They maintain instead that all resident groups in the country are indigenous, in part because they do not wish to grant primacy of one group over another. It is extremely difficult, therefore, to obtain reliable census data broken down along tribal affiliation or ethnic group membership lines. Estimates of the number of indigenous peoples in Africa thus range from 25 million to as many as 350 million (for one estimate of the number of indigenous people in Africa and elsewhere, see Table 9.2). A very small percentage of Africa's people are or were hunter–gatherers (see Table 9.3). These groups are often considered aboriginal or indigenous because of their long-term occupancy of the regions in which they live). Hunter-gatherers, pastoralists, and farmers in Africa are flexible, resilient and innovative in their approaches to solving economic and environmental problems. They intentionally enhance biological diversity in order to ensure

Table 9.1. Geographical, population and economic data on African countries

Name of country	Size (km^2) (1992)	Population (1992)	Gross Domestic Product (GDP) ($)
Algeria	2 381 740	26 666 921	54.0 billion
Angola	1 246 700	8 902 076	8.3 billion
Botswana	600 370	1 292 210	3.6 billion
Burkina	274 200	9 653 672	2.9 billion
Cameroon	475 440	12 658 439	11.5 billion
Cape Verde	4 030	398 276	310.0 million
Central African Republic	622 980	3 029 080	1.3 billion
Chad	1 284 000	5 238 908	1.0 billion
Comoros	2 170	493 853	260.0 million
Congo	342 000	2 376 687	2.4 billion
Djibouti	22 000	390 906	340.0 million
Egypt	1 001 450	56 368 950	39.2 billion
Equatorial Guinea	28 050	388 799	156.0 million
Ethiopia	1 221 900	54 270 464	6.6 billion
Gabon	267 670	1 106 355	3.3 billion
The Gambia	11 300	902 089	207.0 million
Ghana	238 540	16 185 351	6.2 billion
Côte d'Ivoire	322 460	13 497 153	10.0 billion
Kenya	582 650	26 164 473	9.7 billion
Lesotho	30 350	1 848 925	420.0 million
Liberia	111 370	2 462 276	988.0 million
Libya	1 759 540	4 484 795	28.9 billion
Madagascar	587 040	12 596 263	2.4 billion
Malawi	118 480	9 605 342	1.9 billion
Maldives	300	234 471	174.0 million
Mali	1 240 000	8 641 178	2.2 billion
Mauritania	1 030 700	2 059 187	1.1 billion
Morocco	446 550	26 708 587	27.3 billion
Mozambique	801 590	15 469 150	1.7 billion
Namibia	824 290	1 574 927	2.0 billion
Niger	1 267 000	8 052 945	2.4 billion
Nigeria	923 770	126 274 589	30.0 billion
Réunion	2 510	626 414	3.37 billion
Rwanda	26 340	8 206 446	2.1 billion
São Tomé and Principé	960	132 338	46.0 million
Senegal	196 190	8 205 058	5.0 billion
Seychelles	455	69 519	350.0 million
Sierra Leone	71 740	4 456 737	1.4 billion
Somalia	637 660	7 235 226	1.7 billion
South Africa	1 221 040	41 688 360	104.0 billion
Sudan	2 505 810	28 305 046	12.1 billion
Swaziland	17 360	913 008	563.0 million
Tanzania	945 090	27 791 552	0.9 billion
Togo	56 790	3 958 863	1.5 billion
Tunisia	163 610	8 445 656	10.9 billion
Uganda	236 040	19 386 104	5.6 billion
Western Sahara	266 000	201 467	60.0 million
Zaire	2 345 410	39 084 400	9.8 billion
Zambia	752 610	8 745 284	4.7 billion
Zimbabwe	390 580	11 033 376	7.1 billion

Source: *The World Factbook* (1992). Washington, DC, US Government.

long term economic survival and ecological sustainability, as can be seen in cases where foragers and pastoralists use fire to promote the growth of desirable plant species and farmers utilize a variety of different types of crops in order to reduce risk and increase the chances of obtaining returns.

Table 9.2. Estimated number of the world's indigenous peoples

Region	Number of groups	Overall population	
Africa	**2 000**	**50 000 000**	
Batwa (Pygmies)		200 000	(7 countries, central)
Bushmen (San)		95 000	(6 countries, southern)
Byle (Somalia)		450	
Hadza (Tanzania)		1 000	
Maasai (Tanzania/Kenya)		500 000	
Tuareg (Tamacheq)		3 000 000	(5 countries, west)
Australia and New Zealand	**100**	**550 000**	
Aboriginals		300 000	
Maaori (New Zealand)		250 000	
China and Japan	**56**	**67 000 000**	
Ainu (Hokkaido, Japan)		26 000	
Shui (Guizhou, China)		280 000	
Former Soviet Union	**135**	**40 000 000**	
Saami (Russia)		65 000	
Latin America and the Caribbean	**800**	**40 000 000**	
Ache (Paraguay)		400	
Mapuche (Chile)		600 000	
Miskito (Nicaragua)		75 000	
Yanonami (Brazil, Venezuela)		20 000	
North America	**250**	**3 500 000**	
Indians (Canada)		1 500 000	(633 bands)
Indians (United States)		2 000 000	(515 tribes)
The Pacific	**750**	**2 000 000**	
Papuans (New Guinea)		1 300 000	
South Asia	**700**	**70 000 000**	
Adivasis (India)		63 000 000	
Tribals (Bangladesh)		1 200 000	
South-east Asia	**500**	**30 000 000**	
Orang Asli (Malaysia)		71 000	
Penan (Borneo)		20 000	
Thailand Hill Tribes		484 000	
GRAND TOTAL	5 290	357 000 000	

Note: Data obtained from a map entitled 'Earliest residents', *The World Monitor* 6(3):II (1993) as well as Burger (1990); IWGIA (1992); Durning (1992); and Hitchcock (1993).

Table 9.3. Indigenous African populations who are or were hunter–gatherers and researchers who have worked with them

Name of group	Location	Researchers
Koroka, Kwepe, Kwise	Angola	A. de Almeida
Va-Nkwa-Nkala	Angola	S. Souindola
San (Basarwa)	Botswana	L. Marshall, J. Marshall, Kalahari Research Committee (Witwatersrand University), Harvard Kalahari Research Group, A. Barnard, M. Guenther, H.J. Heinz, G. Silberbauer, J. Tanaka, E. Wilsmen, P. Wiessner, University of New Mexico Kalahari Project, S. Kent, H. Vierich, J. Yellen, A. Brooks, P. Motsafi, G. Childers, A. Thoma
Fuga	Ethiopia	W. Shack
Boni	Kenya	D. Stiles
Dahalo	Kenya	D. Stiles
Dorobo (Okiek)	Kenya	C. Chang, M. Ichikawa, C. Kratz, G. Huntingford, R. Blackburn, C. Hobley
Mukogodo	Kenya	G. Worthy, L. Cronk
Waata	Kenya	B. Heine, D. Stiles
Mikea	Madagascar	D. Stiles, B. Kelly
Ovatjimba	Namibia	B.J. Grobelaar, H.R. MacCalman, M. Jacobsohn
San (Bushmen)	Namibia	Marshalls, R. Gordon, T. Widlok, Nyae Nyae Fdn.
Batwa (Pygmies) (e.g. Aka, Efe, Sua, Mbuti)	Ituri Forest, Zaire, Central African Republic, Cameroon, Congo, Rwanda, Gabon	D. Turnbull, M. Ichikawa, J. Hart, T. Hart, R. Harako, G. Morelli, P. Putnam, L. Cavalli-Sforza, D. Wilkie, P. Schebesta, B. Hewlett, S. Bahuchet, H. Guillaum, J. van de Koppoll, T. Tanno, P. Ellison, J. Pedereen, R. Bailey, N. Peacock, H. Bode, I. DeVore, R. Aunger, S. De Zalduondo, H. Terashima, Z. Waehle
Eyle	Somalia	S. Brandt

(*continued over*)

Table 9.3. (*cont.*)

Name of group	Location	Researchers
Kilii	Somalia	D. Stiles
Hadza	Tanzania	J. Woodburn, J. O'Connell, K. Hawkes, E. Ten Raa, N. Blurton-Jones, H. Bunn, L. Bertram, D. Ndagala, K. Tomita, J. Newman, L. Smith, W. McDowell
Kwandu	Zambia	B. Reynolds
Amagili	Zimbabwe	A. Campbell, R. Hitchcock
Doma (VaDema)	Zimbabwe	C. Cutshall, R. Hassler

Note: Data obtained from publications, archival research, files of development agencies and non-government organizations, and African government reports and censuses.

One of the reasons for Africa's successful implementation of community empowerment and participatory development projects is that a wide range of community organizations are involved in self-help, development and conservation activities (see Table 9.4). The organizations often manage resources co-operatively, as is the case among grazing associations in Lesotho and the Horn of Africa and forestry committees in East and West Africa. Numerous communities and individuals in Africa have called for a new approach to development—one which is not socially and environmentally destructive. They argue that they have a right to sustainable development, development which has been defined as that which '. . . meets the needs and aspirations of the present without compromising the ability of future generations to meet their own needs' (World Commission on Environment and Development, 1987: 13). This approach is seen by many individuals and groups in Africa as the only way to overcome the difficulties people are experiencing.

One strategy for promoting sustainable rural development currently being debated is the use of local common property resource management (CPRM). Common property resource systems combine local control of resources with measures to promote sustainable use. More and more communities and non-governmental organizations are arguing in favour of community-based resource management as a sustainable development strategy (Associates in Rural Development, 1992). There are growing numbers of projects and community activities in Africa which are attempting to implement community-based resource management projects that are participatory in their orientations (see Appendix 9.1). Some of these projects have been relatively successful, while others have faced constraints ranging from lack of sufficient resources to the unwillingness of higher-level institutions to decentralize authority to grassroots-level organizations. In the balance of this chapter, we consider some

Table 9.4. Community organizations in Africa involved in self-help, development and conservation activities

Type of organization	Location
Farmers' Association	Widespread
Multipurpose Development Organization	Widespread
Co-operative	Widespread
Pastoral Association	Kenya, Somalia
Grazing Association	Lesotho
Refugee Assocation	Ethiopia, Sudan, Malawi
Women's Organization	Kenya, Swaziland, Tanzania, Zimbabwe, Botswana
Fishermen's Co-operative	Mozambique, Zambia, Botswana
Hunters' Co-operative	South Luangwa, Zambia, Binga, Kyaminyami, Zimbabwe
Village Council	Zimbabwe
Council of Elders	Widespread
Fencing Group	Botswana
Village Development Committee (VDC)	Botswana
Burial Society	Botswana, Swaziland
Water Committee	Widespread
Health Committee	Widespread
Sanitation Committee	Widespread
Conservation Committee	Botswana
Forestry Committee	Widespread

Note: Data obtained from archival and fieldwork and from the following sources: Harrison (1987); Pradervand (1989); Durning (1989, 1992); Hitchcock (1993).

models of community empowerment that have been attempted in parts of southern Africa in an attempt to illustrate some of the strategies that have been employed in the promotion of sustainable development in African environments.

Case 1: The Purros Project in Namibia

An example of a sustainable development programme is being pursued by a Namibian non-governmental organization, Integrated Rural Development and Nature Conservation (IRDNC) and some Himba communities in the Kaokoland region of north-western Namibia. The Purros Project is a multi-pronged small scale development effort aimed at improving the lives of several hundred Himba and Herero pastoralists residing in and around the community of Purros (for discussions of this project, see Owen-Smith and Jacobsohn, 1989; Jacobsohn, 1991).

The Himba and Herero can be described as 'multi-use strategists'. They combine semi-nomadic pastoralism with periodic wage labour and small-scale rural industries such as handicraft manufacture. Like many other indigenous populations in southern Africa, they had extensive environmental knowledge, small group sizes, widespread population distribution, and relatively simple but efficient technology. The use of buffering strategies and social mechanisms to bring pressure on individuals involved in over-exploiting resources were key means of ensuring the long-term sustainability of resources. Communal mechanisms also serve to control exploitative intrusive behaviour of outsiders.

One of the components of the Purros Project is a community game guard system that employs Himba and Herero men to oversee an area of 95 000 square kilometres of rugged mountains and semi-arid sandy plains. The game guards, who are not armed, have been able to stop the killing of desert elephants and black rhinoceros at a cost which is much lower than that of the paramilitary anti-poaching campaigns conducted elsewhere in Africa. The Himba and Herero have been able to do this not only because of the presence of game guards who monitor the area carefully but also because of the social pressure brought to bear on individual community members to conserve resources.

A second component of the Purros Project is ecotourism. A levy of 25 rand (about US$12) per person has been imposed on tourists entering the area, and the money generated is divided among community members as they see fit. Initially the idea was to give the money to male household heads, but the women objected strenuously to this plan. Eventually, it was decided to allocate the money equitably among the various household members.

Tourists are given information as to what they should and should not do in the Purros area. Driving across old campsites—which are considered sacred by the Himba—is not allowed. In addition, tour operators have been told by local community members that the tourists they bring in should make it a point to stop and greet community residents before taking photographs or purchasing handicrafts. They have been given estimates of what fair prices are for the crafts bought from the Himba and Herero and have been made aware of the need to limit the environmental impacts of tourism, including firewood depletion and the leaving of refuse.

The Purros tourist system represents an institutional response to a potential resource and land-use conflict situation. By agreeing to a fixed set of rules and by determining specifically who was eligible for benefits, the Himba and Herero have been able to profit equitably and sustainably from their wildlife and the tourism it attracts without increasing social tensions and environmental degradation. The benefits that have been generated by tourism have been re-invested in community activities.

Photograph 9.2. This Herero woman can be described as a 'multi-use strategist'. (Photo: Daniel Stiles)

Case 2: The J/'hoansi Bushmen (San) Development Programme in Namibia

The J/'hoansi Bushmen Development Programme in north-eastern Namibia is an integrated rural development programme that began in 1981. Initiated originally as a 'cattle fund' to provide J/'hoansi groups with livestock, tools and seeds, it has grown into a multi-faceted development programme which is characterized by close co-operation between a non-governmental organization, the Nyae Nyae Development Foundation (NNDF), and a community-based organization, the Nyae Nyae Farmers' Co-operative (NNFC). A key

Photograph 9.3. J/'hoansi women are actively encouraged to participate in the Nyae Nyae Farmers' Co-operative leadership. (Photo: Daniel Stiles)

feature of the programme is the empowerment of J/'hoansi communities through a bottom-up development approach.

In the early 1980s three J/'hoansi groups elected to leave the settlement of Tjumikui in Eastern Bushmanland and return to their original territories. They did this in part to get away from the oppressive atmosphere of a government-sponsored settlement where social conflict, poverty and dependency were rife. They also did it in order to re-establish their claims to their ancestral lands. By 1986, when the NNFC was formed, there were over a dozen communities residing in their traditional territories and pursuing a mixed economic strategy of foraging, livestock production, gardening and wage labour. As of 1992, there were 35 such communities, each with their own water source and surrounding foraging and grazing areas.

The NNDF has provided technical assistance and funding to the J/'hoansi communities in their efforts to become self-sufficient. Emphasis has been placed on social and economic development along with education and training. The Foundation has used a variety of participatory development strategies along with innovative organizational communication methods. Human resource development has been a primary focus of the programme with literacy, vocational and on-the-job training activities being supplemented with both formal and non-formal education.

The model being used by the Foundation stresses communication and self-determination at all levels. The formation of the NNFC was a result of close consultation with local J/'hoansi, none of whom had any experience with setting up and running representational bodies. Initially, the J/'hoansi had open meetings in which literally hundreds of people participated in the traditional style of consensus-based decision making. Later, the communities began to delegate some of the responsibility for attending meetings to specific individuals. Elections were held, and two representatives were chosen from each of the communities to take part in the co-operative meetings. Women's participation in the NNFC leadership was encouraged actively by the J/'hoansi.

Some of the activities of the NNFC include seeking control over the land and resources of Eastern Bushmanland, promoting development activities and taking part in land use and environmental planning with government agencies, NGOs and people in the private sector. The NNFC has acted as a corporate body in seeking to convince outsiders who have moved into the area with their cattle to move elsewhere. The Co-operative has also collaborated with the Foundation in seeking ways to promote better resource management. It has taken part in studies (e.g. a conservation and development planning exercise conducted in early 1991) which have led to the formulation of recommendations on land use. In addition, the NNFC has sought to draw attention to issues relating to land tenure and resource rights at national and regional conferences (e.g. the National Conference on Land Reform and the Land Question held in Windhoek in June–July 1991).

The NNFC operates as an independent body which considers public policy matters at its meetings. It undertakes trips to all the communities in Eastern Bushmanland in order to listen to the concerns of local people, and it provides them with information. It has made people aware of political, economic and environmental issues. It has its own bank account and it runs its own co-operative shop and handicraft purchases operation. Over the six years of its existence, the Co-operative has evolved into a flexible, effective organization for internal communication and external representation.

The NNDF and the NNFC together and separately have used a participatory approach to development. No decisions are made without having gauged the opinions of the entire population. In some cases, this has meant that new projects have had to be held in abeyance until such time as all the communities were contacted, a process which is by no means easy in a setting in which 35 different villages are dispersed across a 6300 km^2 area. The advantage of this approach is that once initiatives are agreed upon, they have the full support of the J/'hoansi, who then play key roles in implementation (e.g. in providing community labour).

In the past four years, the NNFC has participated in exchange visits with other groups in communal areas of Namibia. The members of the Co-operative have shared their experiences and have sought to learn how others have dealt with development problems. The NNFC is taking part in leadership training and is in the process of expanding its management team and setting up its own administrative staff. It is aiming at playing a leading role in a detailed land use planning exercise for Bushmanland. The NNFC hopes to be able to establish itself as a self-sustaining development organization within the next decade.

DIRECT AND INDIRECT PARTICIPATION

It has become a truism that the failure of many development projects is a result of lack of direct and indirect participation of local people who theoretically are supposed to be beneficiaries. In some cases, development agencies take a 'top-down' approach in which local people are not consulted before, during or after the implementation of the project. In other cases, people may be asked whether they agree with the project goals, but they do not have any say in the ways in which the project is implemented.

The most effective development projects are those which incorporate local people in decision making at every stage of the development process. Consultation alone, however, is insufficient. Local people must play a role in the identification of problems and constraints; they must assist in designing interventions to address those factors; and they must be part of the management of whatever programmes or projects are established.

There are examples of projects in which management authority is ceded over target areas by government agencies to NGOs. This is the case, for example with the Annapurna Conservation Area in Nepal (Wells and Brandon, 1992: 44, 83–85). There are relatively few examples of situations in which governments have allowed local people total control over resource management and development action. Local communities do sometimes get control over specific resources (e.g. grazing in the case of pastoral associations in eastern or southern Africa, or water in the case of irrigation organization in Mexico, Peru or South-east Asia). Governments can assist local communities through passage of enabling legislation, as occurred in the case of Appropriate Authority status granted to District Councils in Zimbabwe under the Parks and Wildlife Act or the establishment of multiple-use areas in Niger and Uganda.

It is interesting to note that many of the conservation projects established for preservation of biodiversity have had more indirect than direct benefits for

Photograph 9.4. In the buffer zone of Amboseli, Maasai are allowed to continue their traditional land use patterns. (Photo: Daniel Stiles)

local people. In cases such as Amboseli in southern Kenya, for example, the idea was to promote conservation and tourism by declaring the area a national park. Local Maasai were supposed to be able to continue their traditional land-use patterns in the buffer zone around the park while at the same time gaining access to economic benefits from tourism. As it has worked out, wildlife has increased, which has encouraged more tourism, but the benefits accruing to most of the Maasai are relatively small from an economic standpoint. Some Maasai have, however, been able to gain full legal title over land.

Many of the benefits of rural development projects that ostensibly are for local people end up going instead to middle-level institutions. This situation can be seen in the case of the Kajiado County Council near Amboseli, for example, and it is also a feature of some of the district councils in the Communal Areas Management Programme for Indigenous Resources (CAMPFIRE) programme in Zimbabwe such as Tsholotsho. In some cases, local leaders divert some of the resources to their own purposes, as has occurred in the case of the Administrative Design for Game Management Areas (ADMADE) Programme in Zambia or in some programmes in Indonesia and South America. Efforts have to be made to ensure that local communities receive direct benefits in exchange for the costs that they bear.

The participatory development and community empowerment models that are most effective are those which not only promote the involvement of local people in decision making but also ensure that those people have complete control over their own resources. This kind of approach is advocated relatively frequently, but rarely put into place in an effective way, on external assistance in the form of funds or technical expertise. Few communities have complete control over all their resources, in part because most states retain the rights to valuable assets such as minerals and timber or cede those rights to private companies in exchange for a portion of the profits.

One of the problems with CAMPFIRE in Zimbabwe and other rural development programmes in southern Africa is that many of the decisions about resource management come from outside the producer community. This can be seen, for example, in those cases where the district councils make suggestions to lower-level institutions as to how they should spend the money obtained from wildlife revenues. It should be stressed however, that some of the people at the local level have begun to lobby hard for greater decision-making power, something that district councils have begun to take greater notice of. Even if the situation in Zimbabwe cannot at present be described as one in which the communities have been empowered, it is not unlikely that the trend is towards increased participation in decision making at the local level.

CONCLUSIONS

Experience in Africa has demonstrated that there are a number of conditions which must exist if sustainable development is to be achieved:

(1) Communities must have control over the means of production, especially land and capital.
(2) Local institutions should be self-governing and should have a significant voice in resource management.
(3) Communities must have decision-making power and authority to undertake projects that they deem necessary.
(4) Projects must be of sufficient small scale to be managed at the local level of the community or on a multi-community level.
(5) Capital-inputs must be such that they do not overwhelm the capacity of the local institutions to cope with them.
(6) The management and administration of the projects should not be overly complex organizationally.
(7) Local institutions should pursue activities that are beneficial to as wide a number of people as possible and that are equitable in terms of distribution of power and resources.
(8) Project identification, design and implementation must be done in such a way that dialogue between local people and development agencies is ongoing, and the discussions should have effect on the directions the project takes.
(9) Natural resource management and governance regimes must take account of diverse and legitimate interests.

(10) There should be means of ensuring that the environment is not overtaxed by the development activities.
(11) Fair, just, and socially acceptable mechanisms for conflict resolution must be available.
(12) The institutions involved in resource management must be willing to impose sanctions if individuals and communities fail to comply with the rules.

People of Africa have called for greater emphasis on sustainable development strategies. If they are to survive, local people have argued, they must have the opportunity to get secure access to resources, including land, labour, capital and development-related information. They must also be allowed to determine the kinds of projects to be implemented. Community consensus should be seen as crucial to project success.

Participatory development strategies are not being employed in many parts of Africa. Using community-based resource management systems, local organizations have been able to promote development without sacrificing the environmental integrity of their regions. Although not all of their activities are successful, African communities are learning important lessons. As they note, the future of their communities—and of the world generally—lies in their children. Only when the needs of these children are taken into consideration will there be really successful socio-economic development and human rights for all.

APPENDIX 9.1: PROJECTS IN COMMUNITY BASED NATURAL RESOURCE MANAGEMENT AND SUSTAINABLE SOCIO-ECONOMIC DEVELOPMENT IN AFRICA

Project	Country	General comments
D'Kar	Botswana	A development trust known as Kuru composed of five communities involved in a wide range of activities
Kedia	Botswana	A wildlife utilization project in a remote community of 700 in the western Central District
Lorolwane	Botswana	A communal grazing area with an estimated 200 households involved in several rural communities involved in wildlife utilization, tourism, and handicrafts projects
Mabutsane	Botswana	A conservation trust involved in a sanctuary and tourism programme
Nata Sanctuary	Botswana	A sanctuary and tourism programme
Dimokika Wildlife Reserve	Congo	Small-scale exploitation of plants for commercial purposes, crafts
Nouable-Ndoki	Congo	A forest conservation project that incorporates local people, forest income generation, training and employment
Goviefe Agodome	Ghana	Community-based agroforestry efforts by a village mobilization squad
Mount Nimba	Guinea, Côte d'Ivoire	Community resource management efforts to preserve buffer zones
Amboseli	Kenya	Some benefits from use of the park by Maasai in the surrounding area
Maasai Mara	Kenya	Use of national park by pastoralists for grazing, ecotourism activities
West Pokot District	Kenya	Tree planting by Pokot communities in Chepareria Division and formation of local institutions
Njoguini, Gitero, Kabati	Kenya	A multi-community self-help water project that provides irrigation and domestic supplies to three villages
Bokong	Lesotho	Local communities involved in a multi-purpose reserve area that includes reserve grazing, a vultury, and tourism

APPENDIX 9.1 (*cont.*)

Project	Country	General comments
Sehlabathebe	Lesotho	A grazing association consisting of lifestock owners who manage an area of rangeland on a communal basis
Lake Malawi	Malawi	Multi-faceted development along the shores of the lake, ranging from fisheries to agroforestry, tourism
Bore Forest	Mali	A multi-community governance project in which local villagers manage a series of local forest areas
Mopti	Mali	Community woodlot activities by local villagers
Bazaruto	Mozambique	Fishermen in Bazaruto Archipelago involved in a game guard system overseeing turtle nests and tourism
Caprivi	Namibia	Two groups in Mafue involved in a community game guard system
Purros	Namibia	Himba and Herero communities involved in tourism, handicrafts, and a community game guard system
Air-Tenere Nature Reserve	Niger	A wide-ranging conservation and resource management programme that also promotes rural development
Guesselbodi	Niger	A natural forest management project that promotes local development and rehabilitation of a degraded area
Virunga	Rwanda	Tourism related to gorillas in which local people get jobs, some benefits
Central Rangelands	Somalia	Establishment of water points and grazing association formation
Zeederburg	South Africa	A small group of Bushmen on a farm in the northern Cape involved in tourism
Richtersveld	South Africa	Nama groups allowed access to a park for grazing and tourism
Kosi Bay	South Africa	A chiefly game reserve with Tonga and Zulu involved in tourism
Piggs Peak	Swaziland	A number of *zenzele* associations (women's self-help groups) engaged in income-generating activities, horticulture, and fish production

(*continued over*)

APPENDIX 9.1 (*cont.*)

Project	Country	General comments
Ngorongoro	Tanzania	Benefits to local communities from tourism, Conservation Area infrastructure development, conservation and institution strengthening
Usambara	Tanzania	East Usambara Mountains Agricultural Mountains Conservation Project includes agriculture, agroforestry, livestock production, and fisheries
Bwindi	Uganda	The Development Through Conservation Project (DTC) includes forest development activities of local farmers and the strengthening of resource management
Ruwenzori	Uganda	Ruwenzori Mountaineering Service (RMS), indigenous NGO, guides hikers
Lupande (ADMADE)	Zambia	A wildlife utilization programme with environmental education and training as well as village scouts employment
CAMPFIRE	Zimbabwe	Wildlife utilization, tourism, and income generating projects in 20 of Zimbabwe's 55 districts; rural development activities and benefits distribution are included in the programme

REFERENCES

Anderson, D. and R. Grove (eds) (1987) *Conservation in Africa: People, Policies and Practice*. Cambridge and New York, Cambridge University Press.

Associates in Rural Development (1992) *Decentralization and Local Autonomy: Conditions for Achieving Sustainable Resource Management*. Washington, DC, United States Agency for International Development, Research and Development Bureau.

Blackwell, J.M., R.N. Goodwillie and R. Webb (1991) *Environment and Development in Africa: Selected Case Studies*. Economic Institute Development Policy Case Series, Analytical Case Studies, No. 6. Washington, DC, The World Bank.

Bonner, R. (1993) *At the Hand of Man: Peril and Hope for Africa's Wildlife*. New York, Alfred A. Knopf.

Brown, L.R. and E.C. Wolf (1985) *Reversing Africa's Decline*. Worldwatch Paper 65. Washington, DC, Worldwatch Institute.

Burger, J. (1987) *Report from the Frontier: The State of the World's Indigenous Peoples*. London, Zed Press.

Burger, J. (1990) *The Gaia Atlas of First Peoples: A Future for the Indigenous World*. New York and London, Anchor Books.

Cernea, M. (1985) *Putting People First: Sociological Variables in Rural Development*. New York, Oxford University Press.

Chambers, R. (1983) *Rural Development: Putting the Last First*. New York, John Wiley.

Durning, A. (1989) *Action at the Grassroots: Fighting Poverty and Environmental Decline*. Worldwatch Paper 88. Washington, DC, Worldwatch Institute.

Durning, A. (1992) *Guardians of the Land: Indigenous Peoples and the Health of the Earth*. Worldwatch Paper 112. Washington, DC, Worldwatch Institute.

Economic Development Institute (ed.) (1987) *Readings in Community Participation*, two volumes. Washington, DC, World Bank.

Harrison, P. (1987) *The Greening of Africa: Breaking through in the Battle for Land and Food*. London, Earthscan.

Hitchcock, R. (1993) Africa and discovery: Human rights, environment and development. *American Indian Culture and Research Journal* 17(1): 129–52.

International Co-operation Administration (1955) *Community Development Review*. Washington, DC, ICA.

International Work Group for Indigenous Affairs (1992) *International Work Group for Indigenous Affairs Yearbook for 1991*. Copenhagen, IWGIA.

Jacobsohn, M. (1991) The crucial link: Conservation and development. In: Cock, J. and E. Kock (eds), *Going Green: People, Politics and the Environment in South Africa*. Cape Town, Oxford University Press, pp. 210–22.

Lasswell, H.D. and A. Holmberg (1966) Toward a general theory of directed value accumulation and institutional development. In Peter, H.W. (ed.), *Comparative Theories of Social Change*. Ann Arbor, Foundation for Research on Human Behavior.

Lawry, S. (1990) Tenure policy toward common property natural resources in Sub-Saharan Africa. *Natural Resources Journal* 30: 403–422.

Midgley, J. (1986) *Community Participation, Social Development, and the State*. London, Methuen.

Office of Technology Assessment (1984) *Africa Tomorrow: Issues in Technology, Agriculture, and Foreign Aid*. Washington, DC, United States Congress.

Owen-Smith, G. and M. Jacobsohn (1989) Involving a local community in wildlife conservation: A pilot project at Purros, Southwestern Kaokoland, SWA/Namibia. *Quagga* 27: 21–8.

Paul, S. (1987) *Community Participation in Development Projects: The World Bank Experience*. Washington, DC, World Bank.

Pradervand, P. (1990) *Listening to Africa: Developing Africa from the Grassroots*. New York, Praeger.

Rosenblum, M. and D. Williamson (1987) *Squandering Eden: Africa at the Edge*. New York, Harcourt, Brace, Jovanovich.

Stiles, D. (1988) Arid land plants for economic development and desertification control. *Desertification Control Bulletin* 17: 18–21.

Stiles, D. (1994) Tribals and trade: a strategy for cultural and ecological survival. *Ambio* 23(2): 106–11.

Timberlake, L. (1988) *Africa in Crisis: The Causes, the Cures of Environmental Bankruptcy*. London, Earthscan.

United States Agency for International Development (1989a) *Natural Resources Management Project*, Volume 1: *Regional Overview*. Washington, DC, USAID.

United States Agency for International Development (1989b) *Natural Resources Management Project*, Volume 2: *Country-Specific Description*. Washington, DC, USAID.

Wells, M. and K. Brandon, with L. Hannah (1992) *People and Parks: Linking Protected Area Management with Local Communities*. Washington, DC, World Bank, WWF and USAID.

World Commission on Environment and Development (1987) *Our Common Future*. Oxford and New York, Oxford University Press.

Part IV

INDIGENOUS KNOWLEDGE

'Indigenous knowledge' has entered the environment–development vocabulary from the discipline of anthropology, and in the transference emphasis on certain aspects has changed. Anthropologists studying socio-cultural systems analyse indigenous knowledge within the context of all the other cultural subsystems, such as religion, social relations, etc. Indigenous knowledge is used in the environment–development field more as a tool and as a method to improve the planning of interventions aimed at improving agricultural productivity, natural resource management and so on.

The chapters that follow treat indigenous knowledge from three different perspectives: the first as a theoretical discourse comparing Western and non-Western views, the second from within an indigenous people's society, and the last from the pragmatic point of how to use indigenous knowledge to improve land management practices. The challenge that remains is to find the modes of effectively communicating the rural peoples' science to the people who are trying to help them improve their lives within sustainable limits, achieve a good translation of the contents of that knowledge, while at the same time ensuring that the knowledge remains in the 'ownership' of the people who produced it.

10 Listening to the People: The Use of Indigenous Knowledge to Curb Environmental Degradation

SABINE HÄUSLER
Institute of Social Studies, The Hague, Netherlands

INTRODUCTION

My experience in participatory resource management stems from working as a forester in Community Forestry projects in the hills of Nepal for four years (Häusler, 1993) and more recently from development studies and research on the theoretical interconnections between women, the environment and sustainable development (Braidotti *et al.*, 1993). I would like to contribute some reflections about work in progress on the interconnections between power and knowledge and the use of indigenous knowledge in improving environmental management within the context of development.

Experiences derived from the practice of environmental management over the last decades have shown that it is imperative to heed people's aspirations, knowledge and social organization for sound local management of natural resources. The provisions laid out in Chapter 12 of *Agenda 21* for people's participation in strategies to combat desertification, the respect for indigenous knowledge, organization and technology and a bottom-up approach are laudable. However, I will argue that if the International Convention on Desertification is to take people's input and the respect for local knowledges, organization and technology seriously, international, regional and national institutions have to undergo significant internal changes. They will have to move away from a top-down towards a responsive approach to environmental management which can account for the great variety, complexity and heterogeneity of local arrangements of natural resource use.

In order to arrive at sustainable natural resource management there is a need for policy makers to look for a new type of development co-operation which provides space for non-Western cultures and alternative conceptions of development, aspirations and priorities of the local land managers on their own terms. Otherwise, the move to take indigenous knowledge into account will only amount to a reification of the present state of environmental

Social Aspects of Sustainable Dryland Management. Edited by Daniel Stiles

management' spruced up with bits and pieces of information from local, non-Western cultures.

At issue for sustainability are relationships of power at all levels, bargaining power of different actors and different forms of knowledge. Long-term sustainable management of natural resources will be hardly possible if the connections between knowledge and power are not considered. What is needed is a politically sensitive way to bring together Western scientific and indigenous types of knowledge in a process of negotiation of the social actors involved in development intervention. Western development experts and their Western-trained counterparts in the South have to learn to become sensible towards their own biases and critically reassess their roles as both products of and vehicles for Western power/knowledge. They must be aware of the profoundly political nature of their work, even though they may see themselves 'just' as technicians.

DEVELOPMENT AND ENVIRONMENTAL MANAGEMENT—WESTERN DISCOURSES OF POWER

In his writings on power Michel Foucault (1980) has developed a new understanding about the nature of power: far from seeing it as something somebody has and others do not, he describes networks of power relations which permeate societies. The locus of power is not the state; the nature of power is always relational, omnipresent and activated in instances where bargaining processes take place. Foucault pointed out the intimate connections between power and what is accounted the status of truth in modern Western societies. Such 'truth' is produced in and disseminated from large modern institutions, in our case the institution of international development. What he sees as most problematic about Western discourses is that the centres of power and the centres of truth are identical. More often than not, Western scientific enquiry is closely bound up with certain vested interests (via funding mechanisms), mostly economic or military. Furthermore, science via its claims to absolute 'truths' plays an important role in what is socially accorded the status of 'truth' and hence shapes the way we think and act.

In recent years a growing body of theory has been dismantling the claims of Western science to value neutrality, objectivity and universal validity (see, for example, Haraway, 1991; Shiva, 1988). Within Foucault's framework, power and knowledge meet and are articulated over time into specific discourses. The development discourse that has taken shape over the last four decades is intimately linked with privileged subjects as enunciators, institutions, techniques, policies and certain practices in the form of development programmes and projects.

In analysing a particular discourse we need to ask questions such as: What is argued? By whom? In whose interest? Who is included/excluded from it? In the case of the development discourse it is development experts and scientists who have over the last four decades devised the solutions to the problems of developing countries based on scientific theories and practices developed in Western countries. 'Development', a complex set of interrelated processes, has been split up into separate disciplines and technical problems for which technical solutions have been devised by Western experts whose activities have significantly contributed to giving shape to the 'Third World' as backward and underdeveloped, in need of reform. Western control and primacy over the directions of political, economic and social change of the former colonies have thus been maintained uninterrupted.

Development as propagated by the industrialized countries of the North to the South has been a process of Westernization, centralization, normalization and standardization of modes of production, forms of government and other organizations, legal frameworks, etc. Development intervention has promoted certain normative (Western) standards of what development is and what it entails. Intervention practice aimed at the control of patterns of local economic and political development has resulted in fundamental changes such as increased commoditization and institutionalization. These have been vehicles of development which have heightened confrontations between different interests and values (Long and van der Ploeg, cited after Long and Long, 1992).

At the core of this process lies a universalist economistic framework of development with an instrumental conception of the human/nature relation, the ideal of urban life styles with high levels of consumption and an image of human agency as *homo economicus*. The universal introduction of the Western industrial, export-led, economic growth model of development, despite some positive changes, has caused considerable damage to local life styles, cultures and—very visibly—to the environment.

THE COLONIAL 'CLASSIC' APPROACH TO SOIL EROSION AND CONSERVATION

The origins of the prevailing approach to environmental management within the development context goes back to colonial times. The colonial 'classic' approach to soil erosion and conservation consisted of four elements:

(1) The problem was defined as an environmental one. Social and political reasons for soil erosion were excluded. If necessary, colonial soil conservation programmes were implemented by force: the colonial state 'protected' the environment from the 'destructive' practices of the local farmer.
(2) Local mismanagement of the environment was seen as being based on lack of knowledge of the farmer, on his laziness, ignorance and apathy. Hence, the farmers needed to be educated; in particular, the practice of shifting cultivation was stigmatized as a major evil. White settlers' practices were labelled as 'using the land', whereas black farmers 'destroyed the land'.

(3) Overpopulation was seen as a major cause for soil erosion. Hence, population programmes were part and parcel of all soil conservation programmes already during colonial times.
(4) Pastoralists and local cultivators were seen as poor because they were not sufficiently involved in the market economy. Modern methods of mechanized agriculture were propagated as the answer to both bring in cash income and to reduce the need for children as labourers.

The practical effect of this colonial soil conservation discourse was a legitimization for the colonial powers to extract maximum agricultural and forest produce from the colonies for their own benefit. Even though the developmental nation state replaced the colonial administrations, the new national bureaucracies often just took over the development of their countries in the colonialists' fashion, advised by Western development experts. The introduction of 'development' and some time later 'economic development' to the Southern countries after independence has wrought considerable destruction on local life styles and environments.

In the case of forest management scientifically trained foresters, both Western, and increasingly after the widespread introduction of Western forestry education also national, took over the management of national forests. In many countries forests were nationalized after independence to ensure scientific forest management. Forest management was geared at timber production for global markets which in turn would produce revenues for national development. The forest ministries and departments 'protected' the forests from local people's 'destructive' practices. In effect, control over the use of natural resources was transferred from the local to the national level.

The development discourse of recent years largely excludes the processes that have led to environmental degradation in the South. These may be summarized as the introduction of industrial, export-led growth policies, cash crop production at the cost of production for local subsistence, deteriorating terms of trade for the Southern countries, currency exchange rate mechanisms, debt services and austerity measures under Structural Adjustment Programmes, the role of transnational timber and other corporations, lack of effective land reforms, and large-scale corruption by national and local elites. As mentioned above, in many cases the introduction of scientific forest management by a national forest service itself has in some cases exacerbated the problems of deforestation. A confrontation between national and local levels over the use of the forest resource was created, and local people lost the previously held control over the local forests. Their traditional 'indigenous' forest management practices became in effect illegal.

In summary, the conventional technical environmental management approach may be broadly characterized by the following elements: it is a-historical in its analysis of the problem of environmental degradation, deals only with its symptoms, places the environment before local people, penalizes these people for actions resulting from 'their own poverty', maintains the facade of technical objectivity and upholds the belief that the Western development model is in the best interest of the country. At the level of political economy it proposes no change (Baker, 1981). As Redclift (1993: 37) has pointed out, the confused and limited way in which environmental problems are often understood hampers their solution. In place of a central concern with sustainability, across the full range of policy, most governments offer only a series of uncoordinated reactive policies. This approach usually matches laudable intentions with contradictory policy implications at the sectoral level.

During the last decades the development discourse has undergone significant changes; with increasing involvement in the practice of environmental management much has been learnt by development experts such as foresters, agriculturalists, water engineers, etc. about the management of natural resources in the development context. The need to 'listen to the people' (Chambers, 1983) and for bottom-up approaches to development have been widely recognized for quite some time. Within forestry professional circles much effort has gone into the development of an approach away from purely technical, timber-oriented strategies to people's need-oriented social/community forestry strategies. The growing input of rural sociologists and anthropologists as well as gender specialists into forestry work was crucial for bringing out certain underlying assumptions and biases about the needs of target populations as assumed by Western and Western-trained foresters. Foresters have recognized the existence of indigenous forest management regimes and the importance of considering them in the implementation of community forestry work. However, these recognitions are only slowly implemented into practice.

Most recently, there has been a growing recognition of quite sophisticated indigenous knowledge systems applied by local people in managing natural resources. This has been a positive move to value indigenous, previously subjugated knowledges and practices. However, as I will argue below, the use of indigenous knowledge within the developmentalist framework is not without problems.

In the case of the Community Forestry Development Project of the government of Nepal a lot of dedicated expert advice, critical thinking, training of staff, institutional build-up, new legal instruments and other provisions have led to improvements during the last two decades. However, the experience in Nepal has shown that even the most sophisticated, integrated, user-group oriented community forestry schemes have not produced the results hoped for. In essence, the reason for this is that despite the fact that serious effort has been put into a more participatory, user group-oriented approach to local forest management, full responsibility for the management of local forests has only reluctantly been handed over to the local users, in most cases, control ultimately remained with the forest department.

Standardized institutional arrangements have been imposed onto highly variable and complex local situations in which informal local organizations have previously existed. Institutional requirements of outside institutions with their bureaucratic procedures and target orientation have taken primacy over the flexibility and responsiveness needed to give space to local conditions and informal institutional arrangements. In many cases, parallel structures to the very informal existing local organizations have been set up by the forest department.

The main problem is that local people are enlisted to 'participate' in projects and programmes steered and controlled by outside experts and institutions with inherently different perspectives of the problems and solutions to these. Lack of faith by professional foresters in the capacity of local groups to manage local resources responsibly has ultimately led to poor performance, with the result that efforts to design and plan projects 'on behalf' of the communities were more or less as much rejected by the latter as earlier, purely top-down approaches.

DIFFERENT APPROACHES TO THE USE OF INDIGENOUS KNOWLEDGE SYSTEMS IN CURBING ENVIRONMENTAL DEGRADATION

The most widespread approach to the use of indigenous knowledge applied by natural scientists, anthropologists and development experts takes certain elements of local/indigenous knowledge and incorporates these into the body of Western expert knowledge. Such knowledge is collected, taken out of context, incorporated into Western scientific knowledge and then disseminated to a wider audience of farmers/local people in a wider geographic context. Even though it may yield important new technical facts, this approach simply reifies existing power relations and the primacy of Western expert knowledge within the developmentalist framework.

The discovery of the rich body of indigenous knowledges must be seen as a positive move because it stems from a closer look at things 'as they are' at the local level. The usual approach of development and government agencies is to design programmes and projects in their offices with needs and goals defined from the top and more or less disregard the actual situation. 'People's participation' usually begins not at the design stage but when the plans are implemented into practice.

Indigenous knowledge is highly location-specific and based on close observation over long periods of time. It is embedded in culturally based value systems, systems of production and consumption and ways of living and relating to the natural environment. These dimensions of indigenous knowledge systems are usually disregarded when indigenous knowledge is put to use in environmental management.

Indigenous forest management systems, where they exist, are usually highly location-specific with (mostly conservative) harvesting regimes. The use of wood and a host of other products is suited to local needs as they arise and not based on Western, standardized harvesting procedures. The local groups managing the forest are often based on informal and flexible neighbourhood arrangements. Therefore it is very difficult to incorporate them as such into government structures which would require formalization, standardization, scientific forest management procedures and legalization.

The second approach that I discern is the 'knowledge systems view' (Marglin in Apffel Marglin and Marglin, 1990). Two types of knowledge—Western scientific and indigenous—are distinguished on the basis of their characteristics:

- Indigenous knowledge is seen as *techne*, which based on experience is personal, particular, intuitive, implicit, indecomposable and orally transmitted.
- Western scientific knowledge, seen as *episteme*, is analytical, impersonal, universal, cerebral, logically deducted from self-evident first principles and transmitted in written form.

This distinction is conceptually useful because it clearly points to the shortcomings of the developmentalist approach described above:

> The expert recognizes the *techne* of the farmer . . . and is happy to learn from the farmer. In return, the expert is more than willing to teach the farmer, to give back in epistemic reworking that which he has elaborated on the basis of the farmer's *techne*. But this is exactly where the trouble begins. The expert can translate back only what he can translate from the farmer's *techne* into his own episteme . . . Translation is always partial and something is invariably lost. What cannot be translated ceases to exist, or exists only as a residue of superstition, ignorance or belief . . . 'Dialogue' becomes appropriation, reduction, and loss (Marglin, 1991).

The systems of knowledge approach presents the nature of indigenous knowledge in a more differentiated manner than the approach described above. It validates the relevance of non-Western cultures and their respective knowledge systems holistically and at the same time recognizes the problem of appropriation of it by Western scientists. In Western and Eastern highly industrialized countries there is also a vast store of indigenous knowledges (see Banuri and Apffel Marglin, 1993). This approach has produced some very informative comparisons and revealing accounts of the interaction between Western scientific and indigenous knowledges (Appfel-Marglin and Marglin, 1990). But conceptually, the two knowledge systems tend to be seen in a dualistic and somewhat oppositional relationship.

The 'knowledge systems view' shares the dualistic notion of Western and indigenous knowledge with social activists like Vandana Shiva of India (Shiva, 1989). Calls are issued by many southern activists for building alternative development strategies solely upon local forms of organization, resources and knowledge outside of institutions like nation states, Western development agencies and transnational corporations. However, in today's world of global markets and economic relations, and large numbers of people living in urban environments, such visions—even though laudable—hardly seem feasible. As Ashis Nandy (1987) has so aptly pointed out, the choice today is no longer between traditionalism and modernity in their pure forms but an enlightened middle way between the two.

Maintaining a strictly dualistic understanding of Western scientific and indigenous knowledges may in practice lead to a revival of populist strategies and a simple reversal of the hierarchy of scientific over indigenous knowledge.

The third approach I want to present here is Norman Long's actor-oriented approach which sees the issue from yet another perspective. The essence of his approach is that its concepts are grounded in the everyday life experiences and understandings of men and women, be they Western development experts, poor peasants in Africa, local government bureaucrats or researchers. His aim is a more thorough treatment of social change and development intervention which provides accounts for life-worlds, strategies and rationalities of different actors in different social arenas on an equal footing. The objective is an elucidation of the actors' own interpretations and strategies and how these interlock through processes of negotiation and accommodation. This entails recognizing the existence of multiple realities and diverse social practices applied by the actors.

Long refrains from making a dualistic distinction between Western scientific and indigenous knowledge. He asserts that the production and transformation of knowledge does not lie in classification *per se* but rather in the processes by which social actors interact, negotiate and accommodate to each others' life-worlds leading either to reinforcement of existing types of knowledge or to the emergence of new forms (Long, in Long and Long, 1992). In his view simple dichotomous distinctions between knowledge systems do not, for example, account for the creativity and experimentation by farmers and their ability to absorb and rework outside ideas. For him knowledge emerges as a product of the interaction between different actors. These processes and their outcomes are shaped to a certain extent by sources of power, authority and legitimization available to the different actors involved.

Long aims at developing a methodology for handling the complex set of relations evolving in interface situations of development intervention that would allow for a more differentiated understanding of how bodies of knowledge shape struggles and negotiations between local groups and intervening parties. Here, intervention is not seen as a linear process of implementing a plan of action but rather as an ongoing transformation process by which knowledge is negotiated and jointly created through social encounters in which certain power dynamics are operating. Intervention is continuously modified by the negotiations and strategies that emerge between the various parties involved. Important is the attempt to understand not only the power struggles between 'outsiders' and 'locals' but also the working of local power structures. Interestingly, the researcher who investigates these processes and his or her objectives and life-world are an intrinsic part of such studies.

For Long, development discourse and action involve a struggle over images of development and 'the good society'. Particular types of intervention must be placed within a broader sociological and historical framework of analysis that identifies the crucial actors, interests, resources, discourses and struggles that are entailed. Planned intervention must then be deconstructed in order to get away theoretically from existing orthodoxies and simplifications concerning the nature and tendencies of structural change. This would lead to a reconceptualization of development intervention as a complex process involving the articulation and reshaping of different life-worlds and understandings of actors.

Long adds to our understanding that all knowledge is partial and based on a particular perspective. He abandons dualistic distinctions between Western and non-Western knowledges, thus implying an equal validation of different forms of knowledge. This position opens up the perspective for sustainable environmental management as a process of negotiation between actors with different worldviews, interests, resources and power on international, national and local levels. This would imply an actor-oriented approach to environmental management with new institutions acting as facilitators in negotiations and conflict resolution between all groups of actors involved.

CONCLUSION

A more differentiated approach to the study of indigenous knowledge systems may help us to refrain from simplistic notions of using indigenous knowledges as a new panacea or the latest fad in development practice, while distracting attention from more fundamental changes necessary for the implementation of sustainable environmental management globally. The effect of this would be a legitimization of 'more of the same' with slight improvements. Fundamental changes in power relations between local, national and international actors are inevitable for any meaningful changes towards the sustainable use of environmental resources in an intimately interconnected global environment. What is at stake is a fundamental questioning of the Western development model itself, and the ways different groups of people dominate each other and the natural environment. Most important of all, a sensitive approach to the interaction between indigenous and Western scientific knowledge about the use of the natural environment has the potential to open up a space for fundamental questioning, negotiations and redefinitions of what 'development' is and what it entails. This is—in essence—an ethical question.

REFERENCES

Apffel Marglin, F. and S.A. Marglin (1990) *Dominating Knowledge. Development, Culture and Resistance.* Oxford, Clarendon Press.

Baker, R. (1981) Land degradation in Kenya: economic or social crisis? In: Gleave, M.B. (ed.), *Societies in Change: Studies of Capitalist Penetration.* Salford University.

Banuri, T. and F. Apffel Marglin (1993) *Who Will Save the Forests? Knowledge, Power and Environmental Destruction.* London, Zed Books.

Braidotti, R., E. Charkiewicz, S. Häusler and S. Wieringa (1993) *Women, the Environment and Sustainable Development—Towards a Theoretical Synthesis.* London, Zed Books.

Chambers, R. (1983) *Rural Development—Putting the Last First.* London, Longman.

Foucault, M. (1980) *Power/Knowledge.* Brighton, The Harvester Press.

Haraway, D. (1991) *Simians, Cyborgs and Women. The Reinvention of Nature.* London, Free Association Books.

Häusler, S. (1993) Community forestry—a critical assessment. The case of Nepal. *The Ecologist* **23** (3), May/June.

Long, N. and A. Long (eds), (1992) *Battlefields of Knowledge.* London, Routledge.

Marglin, S. (1991) *Two Essays on Agriculture and Knowledge.* Harvard University and World Institute of Development Economics Research.

Nandy, A. (1987) Cultural frames for social transformation: a credo. *Alternatives* **XII**: 113–123.

Redclift, M. (1993) Sustainable development and global environmental change—implications of a changing agenda. *Global Environmental Change*, March.

Shiva, V. (1988) *The Violence of the Green Revolution. Ecological Degradation and Political Conflict in Punjab.* London, Zed Books.

Shiva, V. (1989) *Staying Alive—Women, Ecology and Development.* London, Zed Books.

11 Protection of Forests and Other Natural Resources: a View from Central America

ENRIQUE INATOY
Asociación Napguana (KUNA), Panama

INTRODUCTION

For indigenous people, the earth is intimately connected with their indigenous cultures and it is symbolized as 'Mother' because it offers its inhabitants all the resources necessary for their existence and survival. Mother Earth provides forests, rivers, and a diverse range of flora and fauna, many of which are useful for medical or technological purposes, contributing to a better quality of life. For this reason, indigenous people feel a tremendous respect for their Mother Earth, and they try to live harmoniously with nature as an intrinsic part of their being.

After five centuries of lies from Western cultures, combined with the imposition of foreign ideas, our concepts and viewpoints about the true significance of the richness of the earth's natural resources went through a process of change. During this long historical process, the belief was formed that these gifts from Mother Earth were unlimited. Indigenous people were also led to believe that they were an impediment to the development of civilization, and an obstacle to the activities of economic advancement of a country. At the base of this development concept there came the introduction of inappropriate technologies that irresponsibly exploited the land and marine resources, damaging Mother Earth, as well as plundering indigenous peoples' territories. If we, the indigenous people, have resisted and survived in these tropical jungles until the dawning of today, this is due to our relationship with and our respect for Mother Earth. Nevertheless, when the conservation of nature and its diverseness is debated, the territorial rights of indigenous people are often forgotten, or if they are recognized they are treated as a ghost, considered a secondary priority.

The deterioration of Mother Earth is equivalent to the crisis of global cultural diversity. Indigenous peoples live in areas of high biodiversity, and they are confronted as well by many threats against their territorial, cultural

Social Aspects of Sustainable Dryland Management. Edited by Daniel Stiles

and spiritual possessions, and in some areas their very lives are threatened. This long experience with nature and outside exploitation has been interpreted by and incorporated into our social, economic and political systems, in order to define the territorial limits of our natural resources, and to ensure our existence and future development of new generations of indigenous peoples.

INDIGENOUS PEOPLE OF PANAMA

Panama is a narrow natural viaduct that unites North, Central and South America. Because of its geographic position in tropical latitudes, it has developed throughout its evolution a great biological diversity, both terrestrial and marine. This in turn has promoted the development of diverse indigenous cultures in the region, enjoying local natural resources.

In Panama, approximately 55 000 ha of tropical rain forests are destroyed each year, and it is no coincidence that the principal zones of tropical forests that remain in the country are occupied by indigenous peoples, who need them in order to survive. For example, in the Talamanca mountain range in the province of Bocas del Toro live the indigenous groups Bribri (with 500 inhabitants) and the Teribes with 2200 inhabitants. In the central mountain ranges in the provinces of Chiriqui, Veraguas and Coclé live 120 000 Guayamíes, divided into two groups with linguistic differences, the Bugleres and the Ngoberes.

In the north-east of Panama live the Kunas, who are located in three different jungle zones: the indigenous territory of Kuna Yala, a mountainous coastal region on the Pacific coast, in the highlands of Bayano and in the area of Darien. The Kunas have a population of approximately 48 000. In the Darien region two related indigenous nations live alongside the Kunas: the Emberá, with 15 000 inhabitants, and the Wounaan, with 3000 inhabitants.

TRADITIONAL KUNA EXPERIENCE ON THE USE AND MANAGEMENT OF NATURAL RESOURCES

We Kunas are originally from the high mountainous region of Darien. However, during the era of the conquista and colonization, our forefathers were persecuted by the Spaniards, and they were slowly displaced. Over time they situated themselves throughout the islands on the Caribbean coast of Panama, along the coastal regions, and some small groups located themselves in the highlands of Bayano, in Darien, and in the Department of Antioquia, Colombia, each community having its own socio-political and administrative structures.

The Kuna people interpret the use and management of natural resources through religious songs, ritual ceremonies, dances, restrictions (usually food-related), among other traditions, and if the rules and norms are not followed, this is believed to provoke illness and death, and in this way return the energy to nature, as a way of re-establishing the equilibrium. The preservation of natural resources is a basic and extremely important issue for the Kuna people, because in our cultural tradition, nature exists in communion with human beings. When our children are born we bury their umbilical cords in the earth, and in this way it can be said that a tree has been planted.

In terms of the mythical-religious conception of Kuna history, a legend recounts how the first Kuna being descended to earth:

> I arrived on Mother Earth. I was received by many brothers — one of them was the air . . . this brother did not want for me to sweat, and so he brought me a great fan to refresh me. Another brother offered me fish, lobster, turtles, etc., and this brother is the sea. Another said to me: 'I see that you're dirty, and I want to clean you'; this brother was water. Still another said to me: 'look at the mountains, the forests . . .'; this is the sun. I am 'el hombre de Igua Uala' [the man of the almond tree].

Within this mythical-religious conception is embodied the way in which the Kuna people think about natural resources found in their territories. Our great Neles (people with a special capacity for communicating with extrasensory spirits) have been orally transmitting from generation to generation our respect for nature and its protection. These natural areas contain a great biodiversity of species of plants and animals. Many of them are used as food, in the construction of our homes, as medicine, and in the making of clothing, among other uses.

We Kunas call these natural safe-havens for wild plant and animal life Kalu (Kunas imagine this as a community well guarded with poisoned arrows). The Kalus are seen as sacred places, inhabited by a group of natural species that are facing the threat of extinction, like the Harpira Eagle, the Saína and the Macho del Monte (two types of wild pig), the panther, and others. There are many of these sacred places in our Kuna territories, each designated for the conservation of certain species, examples being Sigli Ibegun Kalu for great turkeys (*Crax rubra)*, Putu Ibegun Kalu for the preservation of certain smaller birds (Perdiz de Arca), and Sur Ibegun Kalu, for the preservation of certain monkeys.

Visits that have not been approved by the supreme chief are not permitted in these protected areas. Everyone obeys this rule, because these areas contain very fragile life forms, the destruction of which might unleash disastrous repercussions such as epidemics. In order to be able to enter one of these zones, one must consult with a Neles, after which a ceremony can be prepared during which a pact is made with the chief of the Kalu. In these natural reserves protected by the Kuna culture, all types of hunting and logging and other exploitation are totally prohibited, as these activities would disturb the tranquillity of the species which live within their boundaries.

In the same way, the natural resources of the sea of Kuna Yala are preserved in a sustainable and pragmatic way; for example, when turtles nest, only one tenth of their eggs are allowed to be gathered and the rest are left for reproduction. Also, it is prohibited to eat turtle meat, and the children are told that in the beginning of creation the turtles were people, and if they were to eat their meat they would be eating human meat. There are several marine and land animals which the Kunas are prohibited to eat. This is one way of instilling in the new generations the idea that they should know how to preserve and protect their land and sea resources for tomorrow.

KUNA SYSTEMS OF AGRICULTURAL PRODUCTION

Indigenous systems of forest use and management involve the combination of four principal activities: agriculture, fishing, hunting and gathering. Cultivated indigenous areas are characterized by their multi-levelled vegetation which impedes erosion and the loss of topsoil due to heavy rains and high temperatures of the tropical rain forests. Another of their characteristics is the diversity of the appropriate distribution of crops which discourages dangerous pests and diseases. Furthermore, many farmers plant their crops taking into consideration the soil conditions and the needs of the particular species they are planting.

In the beginning, the Kuna people practised a production system called *nainu*, meaning parcel farming, which consists of taking advantage of parcels of land which are in their 'resting' phase, for the growing of certain specifically defined crops. This practice takes place after the annual cultivation cycle. In its most traditional form, these practices are integrated with the cultural and socio-economic life of our people.

THE KUNA CONCEPT OF LAND AND TERRITORY

Kunas do not consider themselves owners of the land, nor of the ecosystems or their resources. Rather, the Kunas consider that life is intimately linked with nature, and that it cannot be exploited for individual benefit. On the contrary, Kunas believe that natural resources should be used for collective or community benefit, conserving them for future generations. For indigenous people it is inconceivable to imagine a separation with the 'underwater world', and in practice they have seen the serious and irreversible damage that aquatic over-exploitation and contamination have caused. Because of this, underground natural resources in Kuna territories are seen as the 'natural state inheritance', but the national government does not want to let go of them. Because of this, absurd situations have arisen; for example, the recognition of Kuna peoples' rights to territory but not to the underground natural resources (i.e. minerals) that exist within these territories. In this way the Kuna are in effect lords (owners) of a ghost.

In the last few decades, logging and mining activities have come to be the centre of conflict between the state of Panama and the indigenous nations that live within its borders. Legal statutes do not recognize the rights of indigenous people to their underground resources. The national government has not yet recognized that Panama is a multi-lingual country, and that indigenous people have the right to demand legal control of their territories and of their underground resources, so that they can ensure their preservation, sustainability and the regeneration of the ecosystems.

12 Using Indigenous Knowledge for Sustainable Dryland Management: A Global Perspective

D. MICHAEL WARREN AND B. RAJASEKARAN
CIKARD, Iowa State University, Ames, Iowa, USA

INTRODUCTION

Small-scale farmers and pastoralists constitute by far the majority of the rural population in arid and semi-arid ecological settings. Because of the rapid rates of population growth, small-scale producers are obliged to produce their energy, food, fodder, and income from decreasing supplies of land (Kotschi, 1989). This process often leads to soil degradation, soil erosion, and woodland and pasture destruction in rural parts of developing countries. In many developing countries considerable efforts have been made to conserve soil and water resources. However, a number of dryland management schemes have failed due to inappropriate approaches and inadequate understanding of the socio-economic conditions (Rajasekaran and Martin, 1988; Reij, 1991).

The most important reasons for these failures in soil and water conservation programmes include a dominant top-down approach, and the use of systems which are complicated, expensive, and difficult to maintain, in terms of both labour and capital. Such systems may be difficult to replicate. There is also insufficient training of local users of the system, and a heavy reliance on imported machinery for the construction of conservation works (Reij, 1991). All the more important, attitudes generated by the top-down transfer of technology such as ignorance of local systems by technical assistants, scientific reductionism, and short time horizons have precluded the exploration of local knowledge and decision-making systems and the local organizations involved in the identification of local-level problems and the search for their solutions (Chambers, 1991).

In addition, a number of research station-based dryland technologies have little relevance to small-farm conditions. According to Sanghi and Kerr (1991), the conventional graded bunding system in Andhra Pradesh State, India, is not

Social Aspects of Sustainable Dryland Management. Edited by Daniel Stiles

Photograph 12.1. Research station-based dryland technologies do not always have relevance to small-farm conditions. (Photo: Daniel Stiles)

an appropriate dryland technology under small-scale dryland farming conditions due to the following reasons:

- Continuous bunds leave corners in some fields thus creating the risk of losing the piece of land to the neighbouring farmer.
- Contour farming inconveniences field operations (particularly where multi-row implements are used) and reduces the efficiency of operations (where the desi plough is used) due to repeated cultivation in the same direction.
- Systems based on a central watercourse provide benefit to some farmers at the cost of others with regard to disposal of excess run-off.
- The overall system emphasizes only long-term gains, hence creating an impression that short-term gains are not possible through such measures (Kerr and Sanghi, 1991: 2).

In another case study from south of the Sahara, water-harvesting projects for impoverished nomads have not been successful due to the following reasons (Reijntjes, 1986):

- Cropping does not fit well into the nomadic strategy for survival, in which mobility plays an important role.

- Working the soil is not highly esteemed among nomads.
- Land use is often communal, which complicates issues of individual or group investment in rangeland improvement or tree growing and how the benefits will be shared.
- Because of the high climatic variability, especially in drier areas with red soils, the reliability of the water-harvesting system is low.

On the other hand, several recent publications on indigenous knowledge systems indicate that they form the most viable basis for sustainable approaches to development of workable dryland management and drought mitigation strategies for the future (Ayers, 1994; IFAD, 1993; Mathias-Mundy *et al.*, 1992; Pawluk *et al.*, 1992; Phillips-Howard, 1993; Rajasekaran and Warren, 1990, 1994; Reij, 1991 and 1993; Richards, 1985; Stigter, 1994; Warren, 1991a,b, 1992a,b, 1994a,b; Warren and Rajasekaran, 1994). These studies have begun to influence the attitudes of policy makers and development planners to consider the role of indigenous knowledge in sustainable development (Warren and Rajasekaran, 1993, 1994).

The purpose of this chapter is to explore the value of indigenous knowledge systems and indigenous organizations as they relate to sustainable dryland management and to provide policy guidelines for incorporating indigenous knowledge systems into dryland management programmes. The specific objectives of this chapter are:

(1) To explore the value and importance of indigenous knowledge systems to sustainable development
(2) To describe several case studies on indigenous knowledge as they relate to dryland management
(3) To develop policy suggestions for incorporating indigenous knowledge systems into dryland management programmes and projects.

INDIGENOUS KNOWLEDGE: A VALUABLE NATIONAL RESOURCE

What is indigenous knowledge?

Indigenous knowledge is local knowledge that is unique to a given culture or society (Warren, 1987). Indigenous knowledge is the systematic body of knowledge acquired by local people through the accumulation of experiences, informal experiments, and intimate understanding of their environment in a given culture (Rajasekaran, 1993). According to Haverkort (1994), indigenous knowledge is the actual knowledge of a given population that reflects the experiences based on traditions and includes more recent experiences with modern technologies. Local people, including farmers, landless labourers, women, rural artisans, pastoralists, and cattle rearers, are the custodians of indigenous knowledge systems. These people are well informed about their own situations, their resources, what works and does not work, and how one change might have an impact on other parts of a given system (Butler and Waud, 1990). Rural populations have a variety of formal and informal organizations through which citizens identify, discuss and prioritize community-level problems and seek mechanisms to solve them—often through local-level experimentation and innovations (Warren, 1992a; Rhoades and Bebbington, 1994).

Diversity of indigenous knowledge

Indigenous knowledge systems are:

- Adaptive skills of local people usually derived from many years of experience that have been communicated through 'oral traditions' and learned through family members over generations (Thrupp, 1989)
- Time-tested agricultural and natural resource management practices, which pave the way for sustainable development (Venkatratnam, 1990)
- Strategies and techniques developed by local people to cope with the changes in the socio-cultural and environmental conditions
- Practices that are accumulated by farmers due to constant experimentation and innovation
- Trial and error problem-solving approaches by groups of people with an objective to meet the challenges they face in their local environments (Roling and Engel, 1988)
- Decision-making skills of local people that draw upon the resources they have at hand.

Indigenous knowledge as a valuable resource

Indigenous knowledge is dynamic, changing through indigenous mechanisms of creativity and innovation as well as through contact with other local and international knowledge systems (Warren, 1991a). These knowledge systems may appear simple to outsiders but they represent mechanisms to ensure minimal livelihoods for local people. Indigenous knowledge systems often are elaborate, and they are adapted to local cultural and environmental conditions (Warren, 1987). They are tuned to the needs of local people and the quality and quantity of available resources (Pretty and Sandbrook, 1991). They pertain to various cultural norms, social roles, or physical conditions. Their efficiency lies in the capacity to adapt to changing circumstances. According to Norgaard (1984: 7):

> Traditional knowledge has been viewed as part of a romantic past, as the major obstacle to development, as a necessary starting point, and as a critical compo-

> nent of a cultural alternative to modernization. Only very rarely, however, is traditional knowledge treated as knowledge *per se* in the mainstream of the agricultural development and environmental management literature, as knowledge that contributes to our understanding of agricultural production and the maintenance and use of environmental systems.

Indigenous knowledge includes practical concepts that can be used to facilitate communication among people coming from different backgrounds such as local populations, agricultural researchers and extension workers (Warren, 1991a). Indigenous knowledge helps to ensure that the end-users of development projects are involved in developing technologies appropriate to their needs. By working with and through existing systems, change agents can facilitate the transfer of technology generated through the international research network in order to improve local systems. Indigenous knowledge is cost-effective since it builds on local development efforts, enhancing sustainability and capacity building (Titilola, 1990).

Undermining indigenous knowledge can lead local people to become increasingly dependent on outside expertise (Richards, 1985). The greatest negative consequences of the under-utilization of indigenous knowledge, according to Atte (1992: 30) are the following:

> Loss and non-utilization of indigenous knowledge results in the inefficient allocation of resources and manpower to inappropriate planning strategies which have done little to alleviate rural poverty. With little contact with rural people, planning experts and state functionaries have attempted to implement programmes which do not meet the goals of rural people, or affect the structures and processes that perpetuate rural poverty. Human and natural resources in rural areas have remained inefficiently used or not used at all.

Establishing indigenous knowledge centres and networks

Regional and national indigenous knowledge resource centres have embarked on systematic recording of indigenous knowledge systems for use in development. Three global centres, CIKARD (USA), LEAD (The Netherlands), and CIRAN (The Netherlands), facilitate the establishment of these centres. The two regional centres for Africa and Asia are located in Nigeria and the Philippines, while national centres now exist in Mexico, the Philippines, Indonesia, Ghana, Kenya, Sri Lanka, Brazil, Venezuela, South Africa, Burkina Faso, and Germany. The three global centres provide a partnership relationship with the regional and national indigenous knowledge resource centres by: (1) developing guidelines to establish indigenous knowledge resource centres; (2) co-ordinating the activities of regional and national centres; (3) compiling a list of documents held at CIKARD and making it available to the centres; and (4) developing human resources for the regional/national indigenous knowledge resource centres.

The functions of national indigenous knowledge systems resource centres include:

- Providing a national data management function where published and unpublished information on indigenous knowledge is systematically maintained for use by development practitioners and local communities
- Designing training materials on the methodologies for recording indigenous knowledge systems for use in national training institutes and universities (Warren and Rajasekaran, 1994)
- Establishing a link between the citizens of a country who are the originators of indigenous knowledge and the development community.

A complementary networking system is being established through the efforts of Anil Gupta at the Indian Institute of Management, Ahmedabad, India. A newsletter, *Honey Bee*, now published quarterly and translated into a growing number of languages within India and in other countries, documents indigenous innovations, many of which are found in arid and semi-arid ecological zones (Gupta, 1989).

The Indigenous Knowledge and Development Monitor is the quarterly newsletter of the growing global network of regional and national indigenous knowledge resource centres. It is published in The Hague at CIRAN, the Centre for International Research and Advisory Networks.

Case studies on indigenous dryland management practices

Indigenous dryland management strategies that are evolved, modified, and adopted by local people are effective starting points while planning and implementing new dryland management programmes. Prior to designing and developing dryland projects, it is essential for planners and practitioners to understand how indigenous dryland management strategies work in complex agro-ecological and socio-cultural environments of the small-scale farmers and pastoralists. A search was conducted at the documentation unit and library of the Centre for Indigenous Knowledge for Agriculture and Rural Development (CIKARD) in order to identify case studies pertaining to indigenous dryland management. The following case studies represent the wide spectrum of indigenous dryland management practices adopted by farmers in Africa and India.

Stone lines in Burkina Faso

The Mossi people of Burkina Faso developed stone bunding for soil and water conservation (Reij, 1991, 1993). The bunds (lines of stones) built over the years reach a metre high, effectively terracing the slopes with relatively little labour input, most of it during the dry season. After a series of droughts in the 1970s, the stone bunds were spontaneously revived and combined with *zay*, or pits, which conserve water and in which organic material is placed to increase soil fertility. At the same time, introduced systems were shunned (Anon., 1993).

The bunds are semi-permeable, allowing some water to pass through. The water that would have otherwise 'run-off' the fields and caused erosion is able to slowly sink into the ground and benefit the crops. Erosion is avoided, and the gradual seeping in of the water to the soil helps to build up soil fertility. In the disastrous drought years of 1983 and 1984, crops grew up on land with bunds, while adjoining fields grew nothing. Four years ago, the UN's International Fund for Agricultural Development (IFAD) gave Burkina Faso funds to spread the idea throughout the country's densely populated central plateau. About 150 villages on the plateau now have stone lines with very positive results (IFAD, 1993). Sorghum yields on the plateau have risen by about 40% in fields with bunds (Anon., 1993).

Indigenous dryland agricultural experimentation in Niger

An agricultural communications project implemented in Niger by the Academy for Educational Development discovered that farmers carry out a wide variety of experiments and systematically exchange the results of their research:

> There is incontrovertible evidence that, drawing upon their own resources, African farmers have always been and continue to be great agricultural innovators and experimenters. Sahelian farmers in particular must constantly cope with rapid climatological changes in order to survive. To do so, they require a continuing supply of locally adapted technologies. The assumption behind the Niger study was that farmers are dynamic actors in the process of meeting their needs. Indeed, there is some evidence that producers' propensity to experiment and innovate is greater in highly diversified and/or stressed environments where extension is poor or nonexistent. Much can therefore be learned from farmers. The Nigerian Sahel served as a natural laboratory in which to observe and analyse their adaptive, adoptive, and communicative behavior (McCorkle and McClure, 1994).

Hausa dryland agriculture in Nigeria

Highly innovative approaches by Hausa dryland farmers have been recorded recently by studies in Nigeria (Warren, 1992a; Phillips-Howard, 1993). Hausa farmers living on the Jos Plateau of central Nigeria have begun the regeneration of soils from wasteland produced from years of tin mining activities. Although these wasteland soils have been regarded as virtually impossible to regenerate, Hausa farmers have done so through the incorporation of processed garbage from the city of Jos as well as traditional fertilizers such as cattle-egret manure. Their efforts are most impressive with a wide variety of crops being produced on the rehabilitated soils. A farmer-to-farmer exchange of these new technologies is another important characteristic of this situation.

Water harvesting systems in India

Considering the amount of rainfall and soil type, the farmers in Andhra Pradesh State, India, have developed the following major types of water harvesting systems (Sanghi and Kerr, 1991):

(1) Individual farm ponds for supplemental irrigation or percolation are observed in areas where rainfall is high (more than 750 mm per annum) and where the existing crops are highly sensitive to moisture stress at critical stages.
(2) Community tanks for regular irrigation or percolation are mainly found in red soil areas under a wide range of rainfall conditions (500–1200 mm per annum). Tanks for percolation purposes are used primarily in areas with red soil with low to medium rainfall.
(3) *Khadins* (earthen embankments across the gullies) for harvesting moisture in the root zone are observed in areas with very low rainfall (less than 500 mm per annum) and deep soils. This system recharges the root zone during the *khariff* season for raising a post-rainy season crop under residual moisture as observed in Jaisalmer and Barmer districts of Rajasthan State.

Photographs 12.2 and 12.3. Contrasting indigenous and modern water-pumping systems in India. The former works, but the latter increases output and efficiency. (Photos: Daniel Stiles)

Photograph 12.3.

Steep slope terraces in Cameroon

Farmers in the Mandara Mountains of Northern Cameroon practise an intricate system of indigenously developed terracing of steep slopes (Reij, 1991). They use household and animal wastes and crop residues for maintaining soil fertility. Manure is spread and worked into the soil, and crops are rotated and intercropped. Trees such as *Acacia albida* and *Khaya senegalensis* are grown on the terraces, with their leaves used as fodder.

Mixed cropping and intercropping in India

In Rayalaseema region of Andhra Pradesh State, India, mixed cropping of *ragi* (finger millet) with groundnuts, chillies, and cotton is practised by farmers on thousands of hectares. If all these *khariff* sowings fail due to drought, safflower is sown as a rabi crop. On the other hand, in Telangana region of the same state, *jowar* (pearl millet) with red gram or green gram with cucumber are sown as mixed crops. If all these crops fail to germinate due to erratic monsoons, castor is raised in August as a late-sequence crop (Venkataratnam, 1990). Hence, sequential as well as mixed cropping helps the resource-poor dryland farmers in managing risk situations and also in meeting their subsistence food needs (Rajasekaran, 1993).

Red gram and groundnuts are grown as intercrops in dryland areas of Tamil Nadu, Gujarat, and Andhra Pradesh states. This practice is well established due to their value as legumes, oilseeds, and fodder for livestock (Venkataratnam, 1990). Farmers in arid regions of northern India sow *bajra* (pearl millet) utilizing the pre-monsoonal showers (Gupta, 1987). Grassia tribes of Gujarat State practise mixed cropping and strip-cropping to enable the different sizes of root length to reach the varied levels of ground water (Shankaran, 1988). Such drought-mitigation strategies protect the mixed as well as strip crops from erratic monsoonal rains. Furthermore, these strategies also check soil erosion and maintain soil fertility.

POLICY INTERVENTIONS

It is clear from these case studies that indigenous knowledge systems are invaluable, diversified, and comprehensive, although this is not always the perception among outsiders (Thurston, 1992). In fact, these systems are often overlooked by Western scientific research and development practitioners because of their oral nature (Warren, 1991a). In most instances, these knowledge systems have never been recorded systematically in written form, and hence they are not easily accessible to agricultural researchers, extension workers, and development practitioners (Warren and Rajasekaran, 1993). Hence, by recording these systems, outsiders can understand better the basis for decision making within a given society. Furthermore, by comparing and contrasting indigenous knowledge systems with the scientific technologies generated through international and national-level research centres, it is possible to identify where exogenous technologies can be utilized to improve endogenous systems (Warren, 1987; Rajasekaran *et al.*, 1991).

In many dryland projects there is a gap between scientific knowledge and local knowledge. Dryland management specialists with formal scientific and economic training formulate technologies and policies according to their perceptions of the dryland management problems (Millington, 1992). It is important to understand that local farmers and pastoralists are equally concerned about dryland management, but at the same time have their own set of resource management priorities and limitations. Therefore indigenous knowledge pertaining to local resource management and traditional soil and water conservation mechanisms should form the foundations for future dryland management initiatives and technological interventions.

Policy interventions in the years to come should give priorities to these threatened knowledge systems. The framework for incorporating indigenous knowledge systems into agricultural research and extension organizations has been used as a basis for developing policy recommendations (Rajasekaran, 1993). The policy recommendations are discussed in the following sequential steps: (1) agro-ecological mapping; (2) recording indigenous dryland management practices; (3) identifying research-based dryland management technologies; (4) conducting farmers' experiments; and (5) disseminating farmer-evaluated dryland management technologies.

AGRO-ECOLOGICAL MAPPING

Analysis of the agro-ecosystem of the villages should form the first step of the entire approach. This analysis provides an understanding of the village environment and its physical conditions (Chambers, 1990). Maps and transects are drawn in participatory ways with local persons in order to demarcate the agro-ecological zones. Transects provide an opportunity to characterize the study villages in terms of crops and livestock husbanded, land-use patterns, watersheds, and soil types. Once the physical resources of the villages are clearly understood from the villagers' perspectives, the indigenous categorization of types of farmers uncovers the socio-cultural and economic variables used to distinguish locally important categories of producers. The identification of local organizations and associations is vital for understanding indigenous approaches to identifying, evaluating, and disseminating sustainable dryland management technologies.

RECORDING INDIGENOUS DRYLAND MANAGEMENT PRACTICES

Once the indigenous grouping of farmers is understood, the next step is to record the indigenous knowledge systems of a particular group of farmers with respect to dryland management. In other words, how does a group of farmers try to overcome or adapt to the dryland management problems using their own knowledge? The social scientist in co-ordination with respective disciplinary scientists should provide leadership for recording indigenous knowledge related to dryland management. One of the characteristic features of indigenous dryland management practices is that farmers often adopt new practices as a group rather than individually. Hence, recording the indigenous dryland management practices of various groups of farmers is important. For instance, farmers in low-lying areas of semi-arid regions of Tamil Nadu State, India, adopt the practice of direct sowing of rice in order to conserve water as a collective activity (Rajasekaran, 1993). In another case, farmers who live in low-rainfall areas of Andhra Pradesh State, India, jointly construct *khadins* (earthen embankments) to harvest rain water effectively (Sanghi and Kerr, 1991).

Table 12.1. Differing perceptions of dryland management

Farmers' perceptions	Scientists' perceptions
Bunding on field boundaries	Bunding on contour
Concentration on soil	Concentration on conservation
Short- and long-term advantages	Long-term advantages
Small and gradual investment	Large and one-time investment
Reclamation of gully erosion	Stabilization of gully erosion

Therefore interacting with indigenous associations is one of the appropriate methods of collecting information on indigenous dryland management practices. Numerous methods are now available for recording indigenous knowledge systems (Warren and Rajasekaran, 1994). The group interaction method results in the identification of indigenous dryland management techniques adopted by the group along with their associated problems, traditional beliefs, socio-cultural and economic factors. Group interactions reveal how local perceptions and approaches to dryland management may differ from those of scientists as depicted in Table 12.1 (Sanghi and Kerr, 1991).

IDENTIFYING RESEARCH-BASED DRYLAND MANAGEMENT TECHNOLOGIES

Identification of research-based dryland management technologies should form the next step of sustainable dryland management programmes. Since farmers and scientists each know and understand many things, but have little overlap between their domains of knowledge, farmer–scientist interaction should help both groups learn. Involving research-minded farmers while identifying research-based dryland management technologies is essential.

Selected research-minded farmers should be taken to nearby research stations where they are provided with opportunities to observe various on-station dryland management technologies. They should also be encouraged to ask questions regarding the technologies with a particular reference to environmental adaptability, social acceptability, and cost involved. The International Centre for Tropical Agriculture (CIAT) has recently followed this strategy in order to allow farmers to select a contour barrier to prevent soil erosion (Anon., 1993). Such a participatory approach is a starting point for identifying self-regulating and sustainable dryland management technologies.

FACILITATING FARMER EXPERIMENTATION

Based on the available indigenous and research-based dryland management technologies, the farmer group should be encouraged to decide which one or combination of dryland management technologies they want to test. Research in Africa and India into adaptive responses to drought and famine suggests that farmers' experiments increase in number and complexity after crises (de Schlippe, 1956; Juma, 1987). Farmers may participate with researchers in conducting experiments of the following types:

(1) A farmer group may be interested in incorporating indigenous and research-station based dryland management technologies.
(2) Another farmer group may want to test an indigenous dryland management technique only.
(3) A third farmer group may want to evaluate a research station technique by suitably modifying it to their conditions.

This interactive process enables farmers to obtain a basket of technologies rather than fixed packages.

During the experimentation it is necessary that farmers should be encouraged to use their own evaluation criteria to assess the dryland management technologies. In addition to the farmers' evaluation criteria, the following factors should also be taken into consideration for evaluating the dryland management technologies during farmer experimentation:

(1) Compatibility with agro-ecological conditions
(2) Compatibility with socio-cultural environments
(3) Need for institutional support
(4) Productivity (both land and labour)
(5) Profitability
(6) Risks involved
(7) Need for external resources
(8) Need for institutional support (extension, credit, co-operatives)
(9) Ease of testing by farmers
(10) Labour intensity.

DISSEMINATING FARMER-EVALUATED DRYLAND MANAGEMENT TECHNOLOGIES

The results of farmer experimentation should form the base-line for disseminating the dryland management technologies on a wider scale. During the process of disseminating farmer-evaluated dryland management technologies, the following socio-cultural, economic, and institutional factors should be taken into consideration:

(1) Varying interests of farmer groups and the distribution of benefits among them need to be analysed and addressed.
(2) Although technical assistance may be needed to advise on the construction of sound soil and water conservation structures, farmers should give final approval of the design of structures in their own fields.

(3) A major effort is needed to strengthen indigenous organizations. NGOs should work in co-operation with government programmes to strengthen farmer organizations so that they can facilitate implementation of farmer-evaluated dryland management programmes. This is an important pre-requisite for developing an efficient system for financing soil and water conservation investments (Kerr, 1991). The strengthening of farmer organizations encourages local experimentation relevant to soil and water conservation programmes (Molnar, 1990).
(4) Credit organizations should deal with farmer groups rather than with individual farmers (Kerr, 1991). This reduces the bank's administrative risks and costs. The farmer organization should be entitled to additional loans only after repayment, relying on group pressure to encourage repayment.
(5) There should be a provision for incentives and awards for dryland management project authorities to implement farmer-evaluated cost-effective dryland management programmes. This will encourage innovation and high-quality work.

CONCLUSIONS

A number of top-down dryland management programmes have failed due to inappropriate methods and lack of understanding of the existing socio-cultural and economic conditions. Indigenous dryland management strategies evolved, modified, and adopted by local people are often sustainable in the complex agro-ecological, socio-cultural, and economic conditions of small-scale and marginal farmers and pastoralists of the developing world. However, these strategies are not easily accessible to dryland policy makers and scientists since they are not recorded in written form. Hence, future policies on dryland management should consider and incorporate these threatened knowledge systems. Identifying indigenous dryland management practices, incorporating indigenous and top-down dryland technologies, conducting farmer experiments, strengthening indigenous organizations, and disseminating farmer-evaluated dryland management technologies are policy interventions towards achieving improved productivity and sustainability of dryland management programmes in developing countries.

REFERENCES

Anon. (1993) More food from stone lines. *International Agricultural Development* **13** (4): 11.

Atte, O.D. (1992) Indigenous local knowledge as key to local-level development: possibilities, constraints and planning issues in the context of Africa. *Studies in Technology and Social Change*, No. 20. Ames, Iowa State University, Technology and Social Change Programme.

Ayers, A. (1994) Indigenous soil and water conservation in Djenne, Mali. In: Warren, D.M., D. Brokensha and L.J. Slikkerveer (eds), *Indigenous Knowledge Systems: The Cultural Dimensions of Development*. London, Kegan Paul International (in press).

Butler, L. and J. Waud (1990) Strengthening extension through the concepts of farming systems research and extension (FSR/E) and sustainability. *Journal of Farming Systems Research-Extension* **1**(1): 77–98.

Chambers, R. (1990) Micro-environments unobserved. *IIED Gatekeeper Series* No. 22. London, International Institute of Environment and Development.

Chambers, R. (1991) *Participatory Rural Appraisal Notes on Practical Approaches and Methods*. Unpublished manuscript. Hyderabad, Administrative Staff College of India.

Gupta, A. (1987) Organizing the poor client responsive research system: can tail wag the dog? *IDS Workshop*, London.

Gupta, A. (ed.) (1989–) *Honey Bee*, quarterly newsletter to document indigenous innovations based on people's knowledge systems. Ahmedabad, Indian Institute of Management.

Haverkort, B. (1994) Agricultural development with a focus on local resources: ILEIA's view on indigenous knowledge. In: Warren, D.M., D. Brokensha and L.J. Slikkerveer (eds), *Indigenous Knowledge Systems: The Cultural Dimensions of Development*. London, Kegan Paul International (forthcoming).

IFAD (1993) *Building on Traditions—Conserving Land and Alleviating Poverty*. Video on IFAD's Special Programme for Sub-Saharan Africa. Rome, International Fund for Agricultural Development.

Juma, C. (1987) Ecological complexity and agricultural innovation: the use of indigenous genetic resources in Bungoma, Kenya. Paper presented at *IDS Workshop on Farmers and Agricultural Research: Complementary Methods*, 26–31 July 1987, University of Sussex, Brighton, UK.

Kassas, M.A.F. (1988) Ecology and management of desertification. *Earth '88: Changing Geographic Perspectives*. Washington, DC, National Geographic Society, pp. 198–211.

Kerr, J. (ed.) (1991) *Farmers' Practices and Soil and Water Conservation Programmes: Summary Proceedings of the Workshop on Farmers' Practices and Soil and Water Conservation Programmes held at ICRISAT, Patencheru, India, 19–21 June 1991*.

Kotschi, J. (1989) Ecofarming practices for tropical smallholdings: research and development in technical cooperation. *Working Papers for Rural Development* No. 14. Eschborn, GTZ.

Mathias-Mundy, E., O. Muchena, G. McKiernan and P. Mundy (1992) Indigenous technical knowledge of private tree management: a bibliographic report. *Bibliographies in Technology and Social Change*, No. 7. Ames, Iowa State University, Technology and Social Change Programme.

McCorkle, C.M. and G. McClure (1994) Farmer know-how and communication for technology transfer: CTTA in Niger. In Warren, D.M., D. Brokensha and L.J. Slikkerveer (eds), *Indigenous Knowledge Systems; The Cultural Dimension of Development*. London, Kegan Paul International (in press).

Millington, A.C. (1992) Soil erosion and conservation. In Mannion, A.M. and S.R. Bowlby (eds), *Environmental Issues in the 1990s*. New York, John Wiley.

Molnar, A. (1990) Land tenure issues in watershed development. In Doolette, J.B. and W.B. Magrath (eds), Watershed development in Asia. *World Bank Technical Paper* No. 127. Washington, DC, The World Bank.

Norgaard, R.B. (1984) Traditional agricultural knowledge: past performance, future prospects, and institutional implications. *American Journal of Agricultural Economics* **66**: 874–8.

Pawluk, R., J. Sandor and J. Tabor (1992) The role of indigenous soil knowledge in agricultural development. *Journal of Soil and Water Conservation* 47(4): 298–302.

Phillips-Howard, K.D. (1993) Management of soil fertility among small-scale farmers on the Jos Plateau, Nigeria. Paper presented at the *Pithecanthropus Centennial 1893–1993,* Leiden.

Pretty, J. and R. Sandbrook (1991) Operationalising sustainable development at the community level: primary environmental care. Paper presented at the *DAC Working Party on Development Assistance and the Environment*, London, October 1991.

Rajasekaran, B. (1993) A framework for incorporating indigenous knowledge systems into agricultural research and extension organizations for sustainable agricultural development. *Technology and Social Change Series*, No. 21. Ames, Iowa: Technology and Social Change Programme, Iowa State University.

Rajasekaran, B. and R. Martin (1988) A dissemination and utilization model for dryland agricultural technologies in the developing countries. In: Unger, P., T. Sneed, W. Jordon and R. Jensen (eds), *Challenges in Dryland Agriculture: A Global Perspective.* Amarillo, Texas: Texas Agricultural Experimentation Station, pp. 652–3.

Rajasekaran, B. and D.M. Warren (1990) The role of indigenous knowledge in drought relief activities. *Report: Drought Disaster Mitigation Workshop.* 30 April–2 May 1990. Washington, DC, Office of US Foreign Disaster Assistance.

Rajasekaran, B. and D.M. Warren (1994) *The Role of Indigenous Soil Management Practices: Evidences from South India Journal of Soil and Water Conservation* (forthcoming).

Rajasekaran, B., D.M. Warren and S.C. Babu (1991) Indigenous natural-resource management systems for sustainable agricultural development: a global perspective. *Journal of International Development* 3(4): 387–402.

Reij, C. (1991) Indigenous soil and water conservation in Africa. *IIED Gatekeeper Series* No. 27. London, International Institute of Environment and Development.

Reij, C. (1993) Improving indigenous soil and water conservation techniques: does it work? *Indigenous Knowledge and Development Monitor* 1(1): 11–13.

Reijntjes, C. (1986) Old techniques for new concepts. *ILEIA Newsletter* 5: 12–13.

Reijntjes, C., B. Haverkort and A. Waters-Bayer (1992) *Farming for the Future: An Introduction to Low-External Input and Sustainable Agriculture.* London, Macmillan.

Rhoades, R. and A. Bebbington (1994) Farmers who experiment: an untapped resource for agricultural research and development. In: Warren, D.M., D. Brokensha and L.J. Slikkerveer (eds), *Indigenous Knowledge Systems: The Cultural Dimensions of Development.* London, Kegan Paul International (in press).

Richards, P. (1985) *Indigenous Agricultural Revolution: Ecology and Food Production in West Africa.* London: Hutchinson.

Roling, N. and P. Engel (1988) Indigenous knowledge systems and knowledge management: utilizing indigenous knowledge in institutional knowledge systems. In: Warren, D.M., L.J. Slikkerveer and S.O. Titilola (eds), *Indigenous Knowledge Systems: Implications for Agriculture and International Development.* Studies in Technology and Social Change No. 11. Ames, Iowa: Technology and Social Change Programme, Iowa State University.

Sanghi, N.K. and J. Kerr (1991) The logic of recommended and indigenous soil and moisture conservation practices. Paper presented at the *Workshop on Farmers' Practices and Soil and Water Conservation Programmes.* ICRISAT, Patencheru, India, 19–21 June 1991.

de Schlippe, P. (1956) *Shifting Cultivation in Africa: the Zande System of Agriculture.* London, Routledge and Kegan Paul.

Shankaran, V. (1988) A case study of Bhil Grasia in Western India. Paper presented at *Seventh World Congress on Rural Sociology*, Italy.

Stigter, C.J. (1994) Transfer of indigenous knowledge and protection of the agricultural environment. In: Warren, D.M., D. Brokensha and L.J. Slikkerveer (eds), *Indigenous Knowledge Systems: The Cultural Dimensions of Development*. London, Kegan Paul International (in press).

Thrupp, L.A. (1989) Legitimizing local knowledge: scientized packages or empowerment for Third World people. In Warren, D.M., L.J. Slikkerveer and S.O. Titilola (eds), *Indigenous Knowledge Systems: Implications for Agriculture and International Development*. Studies in Technology and Social Change No. 11. Ames, Iowa State University, pp. 138–53.

Thurston, D. (1992) *Sustainable Practices for Plant Disease Management in Traditional Farming Systems*. Boulder, CO, Westview Press.

Titilola, S.O. (1990) *The Economics of Incorporating Indigenous Knowledge Systems into Agricultural Development: A Model and Analytical Framework*. Studies in Technology and Social Change No. 17. Ames, Iowa State University, Technology and Social Change Programme.

Venkataratnam, L. (1990) Farmers' wisdom for a sustainable agriculture. *Kisanworld* 7(12): 22.

Warren, D.M. (1987) Editor's notes. *CIKARD News* **1**(1): 5.

Warren, D.M. (1991a) The role of indigenous knowledge in facilitating the agricultural extension process. In: Tillmann, H.J., H. Albrecht, M.A. Salas, M. Dhamotharah and E. Gottschalk (eds), *Proceedings of the International Workshop on Agricultural Knowledge Systems and the Role of Extension*. Stuttgart, University of Hohenheim.

Warren, D.M. (1991b) Using indigenous knowledge in agricultural development. *World Bank Discussion Paper* No. 127. Washington, DC, The World Bank.

Warren, D.M. (1992a) *A Preliminary Analysis of Indigenous Soil Classification and Management Systems in Four Ecozones of Nigeria*. Discussion Paper RCMD 92/1. Ibadan, African Resource Centre for Indigenous Knowledge and Resources and Crop Management Division (RCMD), International Institute of Tropical Agriculture.

Warren, D.M. (1992b) *Strengthening Indigenous Nigerian Organizations and Associations for Rural Development: The Case of Ara Community*. Occasional Paper No. 1. Ibadan, African Resource Centre for Indigenous Knowledge.

Warren, D.M. (1994a) Indigenous knowledge, biodiversity conservation and development. In: James, V. (ed.), *The Role of Agriculture in Sustainable Development in Africa*. Westport, CT, Greenwood Publishing (in preparation).

Warren, D.M. (1994b) Indigenous knowledge systems for sustainable agriculture in Africa. In: James, V. (ed.), *The Role of Agriculture in Sustainable Development in Africa*. Westport, CT, Greenwood Publishing (in preparation).

Warren, D.M. and B. Rajasekaran (1993) Putting local knowledge to good use. *International Agricultural Development* **13**(4): 8–10.

Warren, D.M. and B. Rajasekaran (1994) *Utilizing Indigenous Knowledge Systems for Development: A Manual and Guide*. Ames, Iowa State University, Centre for Indigenous Knowledge for Agriculture and Rural Development (in preparation).

Part V

GENDER ISSUES IN NATURAL RESOURCE MANAGEMENT

The feminist movement in the 1970s and 1980s in the North created an awakening of gender issues in developing countries, although in most cases developments are proceeding very differently than they did in the North. There are few demonstrations or court cases related to women's rights in developing countries. Rather, women in more conservative countries are beginning to examine their role and place in society with more open eyes, and customary gender relationships are being questioned. Northern researchers and development assistance workers are also influencing attitudes and beliefs through the publication of studies and by interventions that put emphasis on women and families.

The studies have produced findings that run counter to many preconceived notions about female contributions to household economies and natural resource management involvement. In fact, it appears that in most rural communities the women are at least as important, if not more so, than men in contributing labour, products and income to the family. In spite of the fact that women often spend the most time utilizing natural resources, and they are the ones most affected by land degradation, they have little formal say in making decisions about management questions. Many say this state of affairs must change if sustainable development is to be achieved.

13 Gender and Participation in Environment and Development Projects in the Drylands

MICHAEL M. HOROWITZ AND FOROUZ JOWKAR
Institute for Development Anthropology, Binghamton, USA

INTRODUCTION

Western ignorance of the relevance of women in pastoral and agro-pastoral dryland management continues to contribute to the extraordinarily weak performance of rural development interventions among the poor majority in the world's drylands (see Jowkar *et al.*, 1991). While it is not the only cause of failure (other causes include generally impoverished understanding of the political ecologies of dryland production, the paucity of appropriate technical packages, and the often destructive policies of governments and donor organizations (Horowitz, 1989)), it is certainly among the most salient.

A gender-sensitive approach to dryland development is perhaps even more critical today than it was in 1977 when the first Nairobi Conference on Desertification was held, because whether or not dryland environments in general have continued to deteriorate, it is beyond question that the economic well-being of dryland populations has worsened markedly and the great brunt of that worsening is being borne by women, children and the elderly. On pastoral rangelands, on rainfed agricultural lands, and on riverine flood plains in drylands smallholder households are increasingly less able to sustain themselves *in situ*, and there has been a torrential emigration of labour, primarily young adult male labour, in search of income-generating activities elsewhere. An invariable result is that children, the elderly, and especially women have had to assume disproportionate shares of the burdens of rural production typically without being vested with jural rights over resource management. Let us examine this phenomenon among dryland pastoralists and agropastoralists.

POLITICAL ECOLOGICAL CHANGE ON PASTORAL RANGELANDS

Most livestock development interventions in Africa, the Middle East, and much of Central Asia have been, at best, unimpressive, and many of them

Social Aspects of Sustainable Dryland Management. Edited by Daniel Stiles

actually worsened the conditions they were supposed to repair. In the decade following the Sahelian drought of the late 1960s and early 1970s, more than $600 million was expended for livestock projects in Sub-Saharan Africa alone (Eicher, 1985: 31). Projects were appraised that would fundamentally restrict communal access to rangelands and that sought to increase livestock offtake. Apart from veterinary actions, which are often—although not invariably—appreciated by herders (Crotty, 1980), few, if any, of the range management, hydrologic, genetic, feeding and finishing, and marketing interventions measurably increased production, enhanced producer income, improved the environment, or provided satisfactory returns on investment (Horowitz and Little, 1987).

Livestock development projects achieved their objectives so rarely that during the latter part of the 1980s a general disillusionment with them pervaded many of the major development funding organizations. In 1995, pastoral production systems are threatened almost everywhere, and in the main not by their own actions, population growth, or even changing climate. Herders—women, men, and children—are victimized rather by policies imposed by governments that favour more sedentary peoples and more sessile production systems; by adverse terms of trade between locally produced and imported goods; by a concentration of animal ownership in the hands of a minority of affluent herders and absentee civil servants, merchants, and farmer herd owners; and by inappropriate development interventions implemented with the assistance of bilateral and multilateral financing agencies.

Pastoral women suffer a double bias. In addition to their victimization as pastoralists, a burden they share with pastoral men, they are victimized as women, that is, by androcentric ideologies (both indigenous and Western) that obscure and denigrate their productive and reproductive contributions. Women's lives in pastoral and agro-pastoral societies in Africa, the Middle East, and Central Asia have recently undergone great transformations. Although the precise changes depend in part on the internal stratification of pastoral communities and how these have been incorporated into the larger economic and political processes, the status of women has generally deteriorated. While it is undeniable that jural rights dictating property allocation in subsistence pastoralism have been generally gender discriminatory, women's productive and reproductive contributions have been essential to the viability of nomadic herding.

Men's greater association with herding and formal political processes was balanced by women's domestic responsibilities and by their participation in herd-related tasks, which granted them a measure of autonomy and social status. Cultural values attached to marriage, fecundity, and seniority further legitimized women's power. A decline in the importance of livestock as a production and consumption factor, and changes in household labour and herd composition, have paralleled a steady erosion of female traditional rights and a marked increase in gender-based differential access to property and cash (Talle, 1988).

An appreciation of the changing conditions of pastoral women today must take into account how these production systems are affected by the interconnected processes of economic diversification, marginalization, and social differentiation. In the next section we outline the major threats to the continued viability of pastoral production systems in dry areas of Africa, the Middle East, and Central Asia, and indicate how pastoral women have been affected by these processes.

LOSS OF PASTORAL LAND AND RESTRICTIONS ON PASTORAL MOBILITY

The recurrent ease with which modern states abrogate pastoral land tenure and assign use rights of former grazing lands to non-pastoral peoples reflects not only an anti-nomad ideology but also the fact that political elites rarely are members of herding communities, and often identify with groups with histories of antagonistic relationships with herders. These elites are therefore predisposed to respond negatively to the interests of pastoralists. State attempts to restrict the movement of herders attack the very basis of their survival. As early as the 1920s, Reza Shah instituted a forced sedentarization of pastoralists in Iran (Darling and Farvar, 1972: 678), with a resultant 'sheep mortality of 70 to 80 per cent'. In Eastern Africa, large pastoral land areas were transferred to European settlers (Matampash, 1991: 28). The failure to understand the significance of movement for herd well-being leads to a constant erosion of pastoral property rights, as states seek to reduce the amount of land devoted to herding. In socialist Tanzania, herders were sedentarized in Ujamaa villages from which a large number of animals were grazed on a relatively small area. 'The problem of localizing extensive herds of livestock became quickly apparent, as "villages" became centres of environmental degradation and malnourished animals . . .' (Galaty, 1988: 293).

Many projects funded by bilateral and multilateral donor organizations tried to limit pastoral movements by instituting new water points, veterinary services, and managed grazing techniques for participating herders who would agree to maintain no more than a fixed number of animals within a geographically defined perimeter. Recurrently pastoral peoples are deprived of their land rights in favour of other segments of the population. In general, the following mechanisms are employed:

- Establishing wildlife reserves on rangelands (Århem, 1985; Collett, 1987; Homewood and Rodgers, 1991; Lindsay, 1987; Turton, 1987; Zuppan, 1992)
- Favouring the expansion of rainfed agriculture onto rangelands (Fauck *et al.*, 1983: 51; Horowitz, 1972, 1973, 1975; Horowitz *et al.*, 1983; Horowitz and Little, 1987: 62–3; Ibrahim, 1984: 110–18; Lane and Pretty, 1991) and replacing rainfed and flood-plain agro-halio-pastoral production

systems with large scale irrigation, often on dam-regulated rivers, displacing herding peoples from critical dry-season pastures.

Of these, the encroachment of rainfed and irrigated agriculture most significantly affects the status of women. Sedentarization and women's participation in farming and cash-generating activities have had paradoxical consequences for women regarding their familial relations, gender identity, work load, and status. Shifts in household size and structure of authority, support network, and marriage patterns, lead to status deterioration for the majority of women. In the past, extended households provided a social network where women could share domestic duties. Transhumance tasks, such as erecting and dismantling tents, packing, and moving, were often shared by women, even though each might have been responsible for the daily domestic tasks of her own household.

Where nomadic herding declines in importance, the large extended household loses its utility as the major unit for mobilization of labour and resources

Photograph 13.1. Women's domestic responsibilities in dryland herding societies include packing up the household and moving, as here with the Gabbra in northern Kenya. (Photo: Daniel Stiles)

(Broch-Due, 1983). Reports on the occurrence of polygamous households are inconsistent (Bonfiglioli *et al.*, 1988; Marx, 1984; Sörensen and von Bülow, 1990), but it appears that families on settlement schemes are becoming increasingly nucleated in isolated dwellings, while notions of household self-sufficiency and individual privacy are on the increase (Talle, 1990). Physical isolation and nuclearization of families have undermined the social bases for women's collaboration (Broch-Due, 1983; Sörensen and von Bülow, 1990; Talle, 1988; Watson *et al.*, 1988). Pressed by the need for cash, women are confronted with increasing work loads exactly when the institutional bases for exchanges of service and assistance are being withdrawn and when the ideals of housewifely duties and dexterous performance of feminine domestic chores are gradually gaining prominence as a source of self-esteem among sedentary women (Ensminger, 1987).

Where pastoral women become involved in commercial cropping, they may also face marketing difficulties. Where women are restricted by their reproductive tasks, and possibly by a linguistic barrier and moral codes that constrain female mobility, they have limited opportunities for contacts outside of their households (el-Bushra, 1986). Detached from market sites, they must rely on their male relatives to sell their products, which affects their control over earnings.

Discrimination against women is related not only to legal measures favouring men but is also rooted in androcentric indigenous interpretations of customary rights to land and its products. Women may resist by forming spontaneous collective action groups to counter discrimination and male exclusivity. In Kenya, under both colonial and national governments, gender-discriminatory land-tenure laws increased the intra-household struggle between men and women over access to resources. Using familial kinship idioms, men interpreted to their own benefit various state laws that legitimated individuation of land ownership, without commensurately remunerating women's labour. Most of the information about this process is from agricultural rather than pastoral regions. In the coffee-farming Murang'a district, women resisted their growing subordination by participating in voluntary collective action groups to counter male solidarity based on kinship ideology (Mackenzie, 1990).

Unequal access to farm land may be compensated for by other cash-generating survival activities. Pastoral women are active in handicrafts, sewing, charcoal making, beer brewing, wood and herb collecting, and prostitution (Brown, 1983; Hjort af Ornäs, 1990; Hogg, 1986). As extensions of feminine traditional and domestic tasks, women's income-producing activities require little investment for training, and may create a margin of manoeuvre for their individual independence and freedom of choice when traditional social relations of support are slowly eroding as pastoralism ceases to be an available economic activity. With access to cash-generating activities, it is not surprising to witness a higher rate of divorce among these women (Adu Bobie,

1981; Quoohon, 1985). However, in the larger arena of the labour market and socio-economic hierarchies, where men are generally favoured in terms of prestigious jobs and higher salaries, women's income-generating activities bring little social status (Brown, 1983; Hogg, 1986). These sources of income and empowerment for women and their dependents also pose additional time constraints on their already heavy daily domestic labour load (Lewando-Hundt, 1984; Merryman, 1984), and rarely allow women to accumulate enough capital to rebuild their herd. Among the Samburu, older widows, who can no longer invoke kinship relations to borrow livestock for herd reconstruction, migrate to the town of Isiolo in search of employment, where a few have become successful traders and acquired animals (Hjort af Ornäs, 1989).

Where group ranches have been established they have generally not been successful. Grandin's (1986: 9) assessment of Kenya group ranches is that they 'have not led to destocking, improved grazing management, or increased commercial livestock production . . . Rather, tenure insecurity has fostered increased individuation of production, with concomitant loss of labour economies and increased tensions within the community'. Ownership of group and individual ranches was not granted to women. Gender-discriminatory registration laws contributed to greater age and gender antagonism, with women systematically excluded from any legal claim to landholding.

Among the Maasai, group ranches were part of economic policies that transformed the household structure and gender relations of authority. Women, who in their roles as daughters, wives, and mothers had some claim over animals and participated in processing and distributing dairy products, were slowly disenfranchised under new legal codes and adopted the status of unpaid workers taking care of their husbands' livestock (Kipuri, 1989; Talle, 1988). With the heavy outmigration of men from the pastoral regions, women are increasingly assuming the burdens of livestock management but without the jural authority over land use that ranch membership conveys.

Rangeland privatization in north-east Somalia began about 30 years ago, and has rapidly accelerated in the decade of the 1980s. Today relatively little communal land remains in Erigavo District, with the bulk of social costs for this privatization borne by pastoral women, who are responsible for day-to-day herd management. Women were queried as to the benefits and costs of the process of converting communal lands to private use:

> The most frequently stated benefit of range privatization was that the pastoralists no longer had to move away from their home areas with their livestock . . . That is, both the period and frequency of absence had diminished. Significantly, none of the women listed an increased access to educational and health facilities as being an advantage, in contrast to the beliefs of many Regional Government officials.

On the other side of the balance sheet the pastoral women listed the costs of range privatization as being:

- An increased rate of soil erosion and vegetation loss
- An increased difficulty in finding pasture for their livestock particularly late in the *Jilaal* (long dry season)
- The incidence of private range boundary disputes in which a number of women incurred heavy beatings from their neighbours
- The increased difficulty and time involved in gaining access to water points due to the need to circumnavigate the perceived (and commonly invisible) boundaries of the private range areas of others
- An increased incidence of animal diseases (some of which they had never seen before) due to livestock crowding and the limitation of access to fresh pastures
- The increased stress of having to herd their flocks continually and personally so that they did not cross over their neighbour's unclear boundaries; a requirement which meant that flocks could only rarely be left in the charge of younger children as had traditionally been done.

In addition to livestock management, pastoral women had the responsibilities of domestic production as well as the education and raising of the children. Thus the extra stress and work load associated with their livestock management, which had been imposed as a result of range privatization, was severely felt (J. Prior, 1992: 87–8).

LABOUR MIGRATION AND ECONOMIC DIVERSIFICATION

Since the pastoral household is engaged in a complex series of activities—including farming, dairying, trading, and craft production—the amount of labour that can be devoted to herding is necessarily less than the number of able-bodied persons it contains (even allowing that very young children and the elderly may be pressed into service). Labour migration out of pastoral communities has grave repercussions on resource management, gender relations, and the health and nutrition of those who remain behind. While there is a good deal of data attesting to the labour intensity of pastoral herding, livestock-sector planners have frequently misunderstood the nature of pastoral work loads on dry rangelands, mistakenly assuming that the 'land-extensive . . . production activities in these areas are not also labour intensive' (Little, 1991: 4).

Economic diversification and wage labour migration compound the problem, now reasonably well appreciated for settled agriculturalists, that labour constitutes a scarce resource, and the reduction in available labour for herding may force a reduction in the number of animals that can be maintained by the household, leading to progressive impoverishment.

The failure of pastoral production systems to support an increasing number of poorer households and the subsequent male outmigration has ramifications for women's lives. The degree to which women become involved in herding

Photograph 13.2. Even young children are pressed into service in most herding societies. (Photo: Daniel Stiles)

and herd-related activities, the effect on their labour load, health, and nutritional status, are greatly regulated by their families' wealth and investment strategies, their household developmental cycle, and gender ideology. Families with substantial remittances may hire shepherds to compensate for pastoral labour shortfalls without increasing women's and children's labour loads (Adra, 1983), while those with meagre access to men's income may be obliged to appropriate the uncompensated labour power of women, especially in the dry season.

Male migration that is temporary or rotationally organized may not result in a labour shortage affecting animal husbandry (Morton, 1990). In her discussion of labour migration among Hawazma (Baggara) pastoralists, Michael (1991) notes that men seek wage labour not because herding no longer supports them but in order to earn money to purchase more animals. Because herding among these Sudanese people is carried out mainly by adolescent and young men, the absence of more mature men does not seem to increase the labour burden on those who remain. As they invest remittances in additional livestock, more milk is produced and marketed, which enhances the economic status of Hawazma women.

Among the Samburu of Kenya, where pastoralism has become precarious, combining herding with wage labour is increasingly the norm, with remittances either saved in the bank or invested in real estate and other non-pastoral activities. The decision to migrate is influenced by both the economic status of the sending household and its stage in the development cycle (see Goody, 1962). According to Sperling (1985, 1987), poorer households and those whose livestock losses from drought have made some herding labour redundant, export labour. While some of the remittances may be used for herd reconstitution, most of these funds are devoted to grain purchases.

Households in their early formative years tend not to have surplus labour. Since many of them cannot support themselves exclusively from herding, they often combine with other households at the same stage of development, pooling their separate herds under the care of a few shepherds while the other men leave to find work. Older migrants return to their pastoral households, where, to ensure their being accorded the status appropriate to their age, they abide closely to Samburu rules of etiquette regarding loyalty, generosity, exchange, and moral obligation. Samburu women are less likely to migrate, and those who do leave to participate in food-for-work projects or in odd jobs on the fringes of towns find themselves estranged from the pastoral economy.

The problem of herd management in the absence of men may be compounded where an often rigid androcentric ideology restricts women's mobility, potentially conflicting with the extensiveness of the land involved in pastoral production. The desirability of moving stock to often distant pastures, bringing herdswomen into contact with men from other households, lineages, and even tribes, potentially competes against the ideal of female seclusion, which insists that women remain close to their homes under the supervision of their fathers, husbands, brothers, and sons. In reality, the ideology underlying gender roles among Muslim herders may be negotiable (Antoun, 1968; Abu-Lughod, 1986).

New technologies, such as large cisterns mounted on trucks (Chatty, 1990), may resolve part of the problem by bringing water to the herds so that the animals do not have to be moved to distant water points. Also, in the absence of male labour, herders may opt for less labour-demanding small stock—sheep and goats—in place of camels and cattle (Birks, 1978, 1985; Cole, 1975, 1981). This pattern is especially observed in the Middle East where the demand for mutton has been rising. Israeli Bedouin, whose access to Negev pastures has been severely restricted by the imposition of military bases and agricultural schemes, have begun to feed sheep on residues from commercial food-processing activities (Abu-Rabia, 1987) rather than on natural pastures. With the reduced radius of mobility, enforced by the state, women and girls have become more involved in herding without necessarily endangering family honour (Marx, 1986a,b; Shoup, 1985).

The increase in women's responsibilities for herding has not usually brought about changes in the interpretation of customary rules that would grant them legal control over their households' livestock (Talle, 1988). Salih (1985) says that among the pastoral groups of Omdurman, Sudan, men's involvement in non-herding, cash-generating activities has added to women's domestic work load, including their contribution to livestock-related tasks, but women's control of the herd and its products has not increased. Dairy marketing is now done by male heads of households and through middlemen, which deprives women of their traditional rights over milk animals and participation in sales decisions.

Male labour out-migration and increased female participation in herding have not altered the basic pastoral division of labour where women are charged with domestic chores. Among the Bani Qitab of the Sultanate of Oman (Birks, 1978), male out-migration was encouraged by income-generating opportunities in oil fields. Temporary seasonal migration by men slowly gave way to permanent exodus. Initially, this labour migration was designed to supplement earning from agro-pastoral activity, not to replace them, but cash income took precedence over interest in flocks and herds. Men withdrew from pastoralism, leaving animals under the charge of women, children, and the elderly. Most households have retained only a few animals for hospitality and religious obligation, not for subsistence. Even canned milk is now preferred to fresh milk. Remittances have generally been spent on consumer goods rather than invested in ways to reduce local economic vulnerability.

The declining viability of pastoral economies may encourage the migration of all adult members of the household, as in the case of WoDaaBe women from Niger who migrated with their husbands after the 1969–74 drought (Rupp, 1976). This pattern has continued for the past 15 years. In December 1991, we observed WoDaaBe couples in the Podor region of Senegal: walking from FulBe village to FulBe village, the men sold charms and medications and the women worked as hairdressers. But the opportunities for women to be employed away from the herds are limited, although the data on sedentary and urban occupations of women who used to herd are not quantitatively persuasive. Watson *et al.* (1988, 1989) report that some Turkana women who migrate to urban centres become prostitutes, while others try to eke out a living making charcoal and brewing beer. Many reports (e.g. Broch-Due, 1983; Jakubowska, 1984; Sörensen and von Bülow, 1990) describe how male out-migration and reduced income from pastoralism result in greater individuation of production and a contraction of the kinship arena within which individuals and households might find mutual support.

Men's absences from the pastoral household have affected the status of women and relative autonomy at home and in the larger community. Most reports insist that as men migrate, women's work loads increase and their spheres of autonomy contract, but the data are not entirely consistent on these points. Some of these inconsistencies may result from researchers' interpretations, differences in household wealth, the intensity of adherence to indigenous gender ideologies that subordinate women's access to sources of autonomy, and the degree to which women depend on remittances for their livelihoods.

While a few reports indicate (Lewando-Hundt, 1984) that men's absences from the household have broadened the arena of women's decision making, most note that the decline in transhumant herding and the concomitant male out-migration have increased women's dependency. For example, among the Tugen of Kenya (Kettel, 1986) increased participation of men in migrant wage labour resulted from privatization of land, reinforced by colonial taxation and settlement policies, and this was associated with differential access of men and women to property in livestock, cash, and land, and by a new pattern of power distribution in the household at the expense of women's domestic autonomy. The increased dependence of women on men saw a decline in traditional mechanisms of female solidarity, such as women's age-sets.

Male migration changes the ratio of men to women in home communities; and this is reflected in the marriage pattern and household structure: women marry at later ages; and an increasing number of impoverished, single mothers are found who have little legitimate claim for support from the men's kin groups and lineages. Matrifocality is becoming more common as men opt out of pastoralism. With the pastoral exodus, Turkana seem to be engaging in informal consensual cohabitation, rather than formal marriage, which is denying women entry to broader kin networks (Watson *et al.*, 1988). Without holding the formal status of 'wife', Turkana women cannot legitimately pursue claims to their mates' livestock or milk (Broch-Due, 1990). Brown (1983) reports that integration into the market economy, along with legal measures taken by Botswana that have favoured large landholders and cattle herders, promotes male out-migration, leaving behind many destitute unmarried mothers who receive little support from their mates' kin. Although the government has offered women such social services as education and health, the labour market favours men with better jobs and higher salaries. Women are further disadvantaged by not having access to plough oxen, which adversely affects productivity on their small farms. Finally, Monimart (1989) notes that as men leave the rural areas, women may be forced into marriages with men who remain behind. These are often wealthier men, not forced to emigrate by economic adversity, who are able to accumulate wives in exchange for cash.

THE COMMERCIALIZATION ISSUE

A pastoral production system rarely focuses on a single product, but makes use rather of both 'continuing' (calves, lambs, and kids; milk, butter, and cheese; transport and traction; manure; hair and wool; and occasionally blood) and 'final' (meat; hides and skins) products. All these may have both use and exchange values, although with the incorporation of pastoral economies into the world market system, increasing importance is being accorded to the ways in which these products generate income both for the producing community and for the nation. Although possibilities for expanding the arena of pastoral commercialization often appear attractive to development planners, there may also be substantial costs, especially those associated with greater social differentiation. While some producers clearly profit from sales, others, perhaps the majority, may find themselves further ratcheted into poverty or forced out of livestock rearing altogether.

COMMERCIAL MEAT PRODUCTION

With the exception of oil-exporting nations, countries with large pastoral populations tend also to be among the poorer and less-developed states, countries that are hard pressed to improve their external trade balances and to satisfy internal, mainly urban, demands for better and cheaper food. Herds present themselves as part-solutions to the need both to increase high-value exports and to provide urban populations with meat, and various donor-supported projects—such as stratified livestock production and supplemental feeding schemes in West Africa and ranches in East Africa—have been implemented to increase the volume of beef and mutton sold. Attempts at shifting from a subsistence dairy to a commercial meat orientation have met with little success, and have reaffirmed one well-placed observer's conclusion that 'livestock development projects . . . have probably been the least successful subgroup among agricultural projects' (Cernea, 1991: 217). Some of the meat-oriented projects, although ostensibly for the benefit of small-scale herders, in fact exacerbated social inequities.

The shift from subsistence dairying to commercial meat operations fundamentally affects the pastoral community in general and pastoral women in particular. First, there is a recurrent tension between the state's interest in livestock as a source of meat for urban consumption and for export, on the one hand, and that of pastoral communities who derive much of their subsistence directly from their herds, on the other. A successful beef- or mutton-producing operation requires a high survival of male stock; both heifer and bull calves must receive adequate supplies of milk. For a pastoral community, raising bull calves on milk is a luxury to be enjoyed only after human hunger is satisfied, and during the frequent periods of low milk production male stock beyond herd reproductive requirements may be allowed to starve. That is, in the hierarchy of milk consumers, pastoral children in a subsistence dairy operation take precedence over male animals:

> The resolution seems to be a compromise on calf/child needs by favouring the female calves . . . [and] by eliminating a high percentage of male calves. Meadows estimates that 40% of male calves disappear before the age of 12 months throughout northeastern Kenya and Maasai rangeland, compared with only a 5% mortality for female calves (Dyson-Hudson, 1991: 236).

Shifting from subsistence dairying to commercial meat production tends to reduce the size of the pastoral population: '. . . commercialization can lead to displacement of labour from poorer, subsistence-oriented pastoral households

. . . This loss of a work force of course further reduces the productive capacity of poorer households' (Sikana and Kerven, 1991: 24). Especially those with smaller herds settle as farmers or emigrate to other regions. The economic and social costs of rendering part of the pastoral population redundant are rarely taken into account in cost-benefit analyses demonstrating the economic advantages of increased livestock offtake and reduced national dependency on meat imports for urban consumers. A more immediate consequence of the shift out of dairying is its adverse effect on women's normal authority over the management of household milk supplies—how much is allocated to animals, children, other members of the household, guests, and for marketing—which is compromised by the need to feed milk to male calves.

In commercial beef production, in which the male calf becomes a surrogate consumer of milk for the ultimate urban or foreign purchaser of its meat, women lose both the revenue from milk sales and the status attendant on making decisions relating to the family's food supply and household hospitality (Kerven, 1987). In a comment on the final design report for AID's Eastern Senegal Bakel Range and Livestock Project, Hooglund (1977: 4) concluded that were the project to be implemented,

> '[w]omen would be left without an important labour input into the family economy and without control over family resources. As the status of women depends practically on their position in the subsistence system and symbolically upon the number of milking cattle at their disposal . . . , the status of women . . . would suffer.

Among Nigerian FulBe, women were responsible for the direct marketing of milk and milk products, while men were responsible for the indirect marketing of livestock, using professional brokers. Both dairy and animal marketing occur in the subsistence dairying enterprise; in fact, women enter the market more frequently than men. Although women's individual transactions are smaller than are men's, the aggregate sales by women contribute substantially to household income and therefore to women's status. Thus, even if total household income were to increase through an emphasis on beef rather than on dairy production—an often assumed though rarely demonstrated proposition—the relative contribution to that income from women would decline.

Disenfranchised from their traditional rights over milk animals, dairying, and marketing, women in beef production tend to be limited to caring for newborn animals. In stratified production schemes, where reproduction of young stock is the responsibility of pastoral herders who then transfer the animals to stations where they are fattened for sale or export, the labour of pastoral women is appropriated by wealthy ranchers and feedlot operators (Dahl, 1979: 258; Dahl and Hjort, 1979). The profitability of commercial beef production is closely tied to the gender division of labour where responsibility for rearing young stock is assigned to women. Women's labour contribution to

Photograph 13.3. FulBe women in West Africa are responsible for milk and milk products. (Photo: Daniel Stiles)

stratified production schemes is interpreted as part of their feminine role and, therefore, remains devalued and of low visibility, overshadowed by the dominance of men in these enterprises. While women may contribute heavily in the labour-intensive task of caring for the calves, their male relatives control income from sales to ranchers and feedlot operators. Excluded from the male-run animal marketing network, women even lose control over their own livestock as men will first sell off animals that belong to their wives.

Because of their divergent investment interests in the herd in meat-oriented operations, disagreements between spouses on the allocation of milk offtake arise with increasing intensity and frequency (Kelly, 1986; Michael, 1990; Waters-Bayer, 1988; Joekes and Pointing, 1991). Conjugal conflict over animal and human milk needs (Walshe *et al.*, 1991: 22) is most acute during the long Sahelian dry season when milk yields normally plummet, but even in the rainy season there may not be a surplus of milk to allow for much human consumption.

Declines in milk availability force changes in the pastoral diet and in the amount of time women spend on food processing. Although herders always include some non-pastoral foods in their diet, a meat-oriented production system implies an increase in obtaining food from the market. Grain, vegetable oil, tea, and other comestibles become important components of the diet, and these are obtained for cash (Bahhady, 1981; Pouillon, 1990). Pastoral women continue to be responsible for feeding their households, but the erosion of milk and dairy marketing increases their dependence on men to provide them with needed cash. As cereals become more prominent in the diet, women must spend more time in transforming the unhusked grain into meal and in obtaining fuelwood for cooking (Waters-Bayer, 1988). There are some indications that the dietary shift away from dairy produce negatively impacts on the health of both women and children (Loutan and Lamotte, 1984; S. Prior, 1989; Teitelbaum, 1980).

COMMERCIAL DAIRY PRODUCTION

If commercial meat transactions tend to marginalize women, would not commercial milk transactions enhance women's status and control, since the marketing of surplus milk and dairy products is normally within the women's domain? Although there is an interesting literature dealing with dairying in Asia (e.g. Bradburd, 1990; Bishop, 1989; Salzman, 1987), the Middle East (e.g. Chatty, 1978; Lancaster, 1980), and Africa (e.g. Ezeomah, 1985; Holden *et al.*, 1990; Massae, 1990; Rodriguez, 1987), the data are neither numerous nor conclusive. The thinness of the data on the African dairy trade is attributed by Little (1993) to its gendered nature. 'Without males in the portrait, dairy trade was [to many observers] uninteresting, unimportant, and thus, unlikely to be portrayed.' Little also notes that much of the successful dairy marketing takes place in periurban regions: 'The perishability of dairy products necessitates a distance usually of no more than 40 to 50 kms from a market town . . ., and this is the spatial band that has been ignored by researchers until recently'. The inconclusiveness of the data is due in part to the fact that commercial milk production has not affected women uniformly; the potential benefits accruing to women from dairy marketing are mediated by a host of variables, such as location and distance to markets, government politics, economic status, and such logistical obstacles as inadequate infrastructure and refrigeration.

Where milk production generates a surplus above household and herd consumption requirements, women may benefit by taking it to local markets and to small-town processing centres (Herren, 1990; Kerven, 1987). Michael (1990), reporting on production and marketing of milk and dairy products in the Sudan, states that Baggara women participate in all stages of this economic activity, and that their incomes constitute two-thirds of total annual household budgets. The incomes generated are controlled by women, and often go towards the purchase of new stock or animal feed. Women's marketing links to various private and state cheese factories are facilitated by government pick-up posts and by a chain of middle women who sell the milk to retail sellers. Little (1989) documents the central role played by Somali women in camel milk trade in both large urban areas and small settlements. Pastoral women serving consumers' needs in small encampments may also engage in milk transaction in urban areas, where markets are dominated primarily by older local women traders who act as intermediaries between producers and consumers. The disaggregated nature of the milk market, the relative durability of soured milk, and the low initial capital investment facilitate the participation of distant pastoralist women in milk marketing, which allows them a degree of cash autonomy.

However, the potential gain from commercial milk production has not always been translated into financial empowerment for women. What seems to be a source of status and economic autonomy in subsistence herding may be co-opted by development policies that favour capital-intensive dairy operations using powdered or concentrated milk imports from Western countries.

Waters-Bayer (1985) reports that in Nigeria, where semi-sedentary Fulani women have been in charge of milk processing and distribution, dairy development efforts have concentrated on high-technology operations that process non-indigenous dairy products for urban consumers. Hindered by low milk output and discouraged by low prices offered by large dairy plants, Fulani women rarely are willing to sell their milk to non-local markets. Absence of infrastructure and preservation technology has so far prevented most women's direct access to urban consumers. In rural markets fermentation techniques adopted by women solve the danger of contamination, and also combat the problem of lactose intolerance common in sub-humid areas.

Where the market response to pastoral products is positive, women's gain may further be differentiated by their class position. For example, the integration of the Komachi herders into the national economy of Iran was encouraged by the rising demand for mutton in excess of the national supply, by the population expansion, and by the government-subsidized wheat prices. Wealthy households' economic privileges were maintained through contractual control over the labour of poor salaried shepherds. Access to market was facilitated through middlemen who, by expanding credit to herders, controlled animal market exchanges. Large herd owners also had an opportunity to engage in dairying, a work done primarily by women. While rich herders did not participate directly in herding activities, spending their time in establishing business ties, their wives engaged themselves in milk and dairy marketing. The wives of poor herders were used, through informal exchanges of labour and reward, to assist rich women in their processing activities (Bradburd, 1990).

Because of very large local and regional variations, and because of the relative paucity of available information, it is clearly imprudent to draw general conclusions about the effects on women of increasing commercialization of pastoral produce beyond suggesting that as the scale of operation increases, control over the revenue from exchanges of meat, dairy, wool, skins, and hides shifts from women to men, and from producing households and local traders to powerful and remote corporations. There are entrepreneurial opportunities for women in these trades, but the data indicate that considerably more research is needed if larger-scale externally funded actions designed to benefit pastoral women are to have any chance of sustainable success.

CONCLUSION

In the general impoverishment of pastoral communities, the social losses borne by women exceed those borne by men. As herds become too small to support households, men may migrate for wage labour, and while their self-image as failed pastoralists may be severe, they are still able to earn an income. The status of women depended on their critical contributions to livestock husbandry, and on their responsibility for distribution and marketing of dairy produce. Pastoral women rarely find their labour to be valued in non-pastoral arenas, yet development interventions have largely ignored them. Even where women are left by emigrant men with the bulk of herd management responsibilities, planners have tended to act as if they were economically irrelevant, and projects are rarely informed by changing gender relations of production. This bias towards seeing men even where they are clearly absent, while ignoring present and hardworking women, is probably reinforced by livestock department and project staffing which are almost invariably 100% male.

In the past, pastoral studies had an almost exclusively male emphasis, focusing on jural responsibility for herd management. Far less attention was paid to women's activities regarding milking, dairying, food processing and distribution, small stock managing, animal health actions, and marketing. In the absence of detailed and accessible information about pastoral women, it is little wonder that they were slighted in development interventions that were directed by persons who frequently embraced an anti-nomad attitude. The new literature attests to the significance and diversity of women's economic contributions, especially as modern pastoral communities find it increasingly impossible to sustain themselves as subsistence herders and seek—or are forced—to diversify their productive activities in order to survive. Among the recurrent events that are transforming gender relations of pastoral production are privatization or restricted access to strategic resources of rangelands and water; increasingly adverse terms of trade between pastoral and non-pastoral produce; economic diversification and wage labour migration, sometimes for both men and women but more commonly for men alone; encroachment on rangelands of game parks and agriculture, particularly large-scale irrigated and mechanized agricultural schemes, forcing livestock more intensively to exploit the only lands still available for them; closing of national frontiers to herders from other countries; government efforts to force herders to sedentarize or otherwise to constrain their mobility, to increase offtake, and to convert from dairy-based to meat-based enterprises. Rapid commoditization of the factors of pastoral production, in combination with the events listed above, weakens the ability of many herders to withstand periodic adversities of climate and disease.

The consequences of these transformations on women are frequently deleterious. When poverty and/or government policy forces men to migrate in search of income away from the pastoral area, women assume men's labour burdens in addition to their own, and frequently endure restrictions on their abilities to accompany transhumance; when poverty and/or policy forces sedentarization and a shift from herding to farming, women must till the land, but they are discriminated against in regard to both land rights and access to markets; when dairying gives way to a concentration on beef cattle, women must care for the calves but no longer control the income derived from their labour, and theirs and their children's nutritional well-being is threatened. With the move to privatization, women's property rights and attendant status and economic autonomy suffer, as the state tends to acknowledge only male claims of ownership. Privatization also favours an atomization of labour groups, and so extended household compounds within which labour is often pooled are replaced by nuclear households.

Our understanding of the effects of these transformations on women is not firmly based across the board. While the mechanisms of pastoral and agropastoral production have been well described in a large number of cases presented mainly by anthropologists and geographers, feminist theory has only recently informed on pastoral studies, and in-depth studies of pastoral women are relatively few and not evenly distributed throughout pastoral regions of Africa, the Middle East, and Central Asia. Planners must understand what is known about pastoral women, and also recognize the tentative and incomplete nature of the knowledge. Since participatory development cannot proceed without sound appreciation of our partners' social, economic, and cultural characteristics, it is imperative for these appreciations to be advanced and deepened by new research. The following list suggests the kinds of inquiry needed:

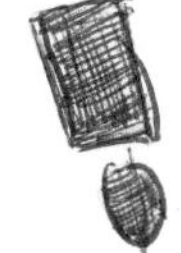

(1) What are the specific effects on pastoral women of the alienation of communal rangelands for agriculture, game parks, ranches, and other forms of private use, especially where women are not permitted to own land in their own name?

(2) What is the impact of social and ethnic conflict and warfare, as in Afghanistan, Somalia, Sudan, and Iran, on pastoral women? Not uncommonly, pastoral men become involved in these conflicts as combatants, often leaving behind unprotected women, children, and the elderly to herd animals on landmined pastures. How women cope with war and brutality in these situations has not been investigated, nor is there a solid body of material on pastoral women in refugee camps.

(3) What is the relationship between women and the processing and marketing of hides, skins, hair, and wool? It is remarkable how few solid data and analyses are available on this topic, and yet anecdotal material suggests that at least with wool and hair women figure prominently. This should be an important area of investigation because of the possibilities for small-scale entrepreneurial activities focused on women.

(4) How are pastoral societies in general and pastoral women in particular coping with the AIDS epidemic? What happens to pastoral societies with

fairly rigid age and gender divisions of labour, when the demographic distribution is abruptly challenged by illness and death? Are there differences in the incidence of AIDS in pastoral and neighbouring agricultural societies?

(5) What labour alternatives exist for pastoral women when herding ceases to be viable? Are pastoral women recruited to particular kinds of employment, such as prostitution, domestic service, tourist industry staffing, craft workers? How does tourism—such as in East Africa and perhaps in the Central Asian republics of the former Soviet Union—affect gender ideologies?

These are only some of the questions that call for field investigations in a number of geographic areas. There are also specific regions where pastoralism is prominent but where scientific knowledge of pastoral production systems and of pastoral women is particularly weak: the Turkic, Mongol, and Farsi-speaking regions of China and the former Soviet Mongolian Peoples' Republic, Tuva ASSR, Khazakhstan, and the new republics of Turkestan. These may be the unique areas of the world where a reinvigorated pastoralism appears to be gaining momentum as state-managed collectivized herding structures are dismantled, and, in the former Soviet countries at least, there appears to be a receptivity to collaborative research between host and foreign social and environmental scientists. The United Nations Environment Programme can play a signal role in these efforts by encouraging research foci on pastoral and agro-pastoral women, and, in so doing, lay the groundwork for effective gender-sensitive development.

Pastoral women are too important to remain neglected by development planners or, worse, to have poorly informed development planning further victimize them. The great contribution of pastoralists—women, children, and men—to the economic lives of their countries is that their animals convert the otherwise unpalatable graze and browse of semi-arid and arid rangelands into useful products (meat, milk, milk products, blood, wool, hair, skin, hides, transport, traction, and manure). Despite their ability to make sustainably productive land areas totalling some 50 million km^2 across the world (Sandford, 1983: 2), much of it having little alternative productive use without very costly and often environmentally disastrous transformations, pastoralists are increasingly among the poorest and least empowered of developing country populations. We hope that the discussions at the UNEP Workshop on 'Listening to the People: Social Aspects of Dryland Management', by highlighting the importance of women's labour in pastoral and agro-pastoral production systems and by demonstrating women's frequent disenfranchisement from cultural and economic sources of autonomy, will alert development planners to the flawed assumptions that lead either to the total neglect of women or to undertakings that discriminate against them, and will facilitate the planning of gender-sensitive development actions that result in enduring improvements in pastoral and agropastoral well-being.

NOTES

1. Dr Horowitz is Director of the Institute for Development Anthropology (IDA) and Professor of Anthropology at State University of New York at Binghamton. Dr Jowkar is a Senior researcher at the institute. They may be contacted at IDA, 99 Collier Street, Binghamton, NY 13902-2207, USA; tel. 1-607-772-624; fax 1-607-773-8993; telex USA 932433 DEVANTHRO; E-Mail FAC398@BINGVAXA.
2. —and there is hardly consensus about the nature, causes, and extent of dryland degradation (Helldén 1990; Warren 1990)—
3. For a discussion of the impacts on gender of shifting from flood-recession to irrigated farming, see Horowitz and Salem-Murdock (1993).
4. This section draws heavily on research conducted by Dr Forouz Jowkar and myself (1992) on behalf of UNIFEM and UNDP.
5. 'Economic diversification' refers to the assumption by members of the household of activities other than maintaining their own herds, including farming, hired shepherding, trading and wage labour. Marginalization 'refers to the compaction of livestock herding on areas of low biological productivity, to the worsening terms of trade between pastoral and nonpastoral produce, and to the pastoralists' political and jural minority'. 'Social differentiation' refers to the growing economic inequality between herders and nonherders and to increasing stratification—by class and gender—within herding communities: '. . . most herders are becoming poorer, despite the relatively high value placed on meat, while a small number of livestock owners from pastoral communities are becoming rich and powerful and perhaps a larger number of rich and powerful men who are not from pastoral communities are becoming livestock owners' (Horowitz and Little, 1987: 61).
6. Because modern states rarely acknowledge pastoral land use as providing inalienable rights, grazing lands are often arbitrarily assigned or granted to agricultural use. In March 1991, the government of Senegal transferred control over a 45 000 ha sylvopastoral zone called Mbegué, which had been sustainably grazed by upwards of 100 000 animals belonging to 6000 FulBe herders, to the Khalifa-General of the Mouride Islamic Brotherhood. Responding to the khalifa's call, disciples clear cut the entire forest within a few weeks, and the land was prepared for cultivation. 'The government's decision to permit the transfer of 45 000 hectares used by agropastoralists to peanut fields controlled by the mouride brotherhood is part of the larger trend toward privatizing communal pastures for individual and sedentary agricultural production. This negatively affects not only the agropastoral livelihood system of the FulBe, but also the environmental sustainability of the entire production system in an area only marginally suited for agriculture' (Schoonmaker Freudenberger, 1991: 2).
7. Efforts to limit pastoral mobility, to privatize rangelands, and to emulate North American and Australian ranching ignore or discount evidence of the superior productivity of 'traditional' pastoral production systems. Studies carried out by the International Livestock Centre for Africa (ILCA), focusing on yields per unit land rather than on yields per unit livestock, show that extensive herding on communal pastures provides the better return. According to Cossins, who has researched Borana herding in southern Ethiopia, their pastoral system 'is very productive; compared with Australian commercial ranches in a similar climatic environment, the Borana produce *nearly four times* as much protein and *six times* as much food energy from each hectare' (1985: 10, emphasis added). The cost of labour and capital invested also favours the Borana system:

> The amount of US dollar investment required to produce one kilogramme of animal protein is 0.14–0.28 for the Borana system, 2.01 for commercial ranches in Laikipia, Kenya, and 1.93–3.89 for Australian ranches. If production of food energy (milk and meat) is used in an indicator, the difference in benefits from pastoral production is even more outstanding (Horowitz and Little, 1987: 71).

Also published by ILCA, similar findings are reported for Botswana by de Ridder and Wagenaar (1984).

8. For a discussion of the gender implications of wool, hair, hides and skin marketing, see Horowitz and Jowkar (1992: 47–9).
9. In Nigeria, the increase in urban milk and dairy consumption led to imports of powdered and concentrated milk. But in 1986 structural adjustment policies resulted in higher import costs. According to Di Domenico and Vabi (1988), the potential gains for pastoral women have been frustrated by inadequate infrastructure.

REFERENCES

Abu-Lughod, L. (1986) *Veiled Sentiments: Honor and Poetry in a Bedouin Society*. Berkeley, CA, University of California Press.

Abu-Rabia, A. (1987) *Stationary Pastoralism among the Negev Bedouin*. Sede Boqer, Ben-Gurion University of the Negev, Jacob Blaustein Institute for Desert Research. (October)

Adra, N. (1983) The impact of male migration on women's roles in agriculture in the Yemen Arab Republic. *Inter-Country Experts Meeting on Women in Food Production*, Amman, Jordan, 22–26 October 1983. Rome, FAO.

Adu Bobie, G.J. (1981) The role of Rendille women. In: *Human Ecology: Consultancy Reports on the Rendille Samburu and the Role of Women/Project 3: Impact of Human Activities and Land Use Practices on Grazing Lands*. Nairobi, UNESCO, pp. 113–161.

Antoun, R.T. (1968) On the modesty of women in Arab Muslim villages: a study in the accommodation of tradition. *American Anthropologist* **70**: 671–97.

Århem, K. (1985) *Pastoral Man in the Garden of Eden: The Maasai of the Ngorongoro Conservation Area, Tanzania*. Uppsala, University of Uppsala.

Bahhady, F.A. (1981) Recent changes in Bedouin systems of livestock production in the Syrian steppe. In: Galaty, J.G., D. Aronson, P.C. Salzman and Amy Chouinard (eds), *The Future of Pastoral Peoples: Proceedings of a Conference Held in Nairobi, Kenya, 4–8 August 1980*. Ottawa, IDRC, pp. 258–66.

Birks, J.S. (1978) The mountain pastoralists of the Sultanate of Oman: reactions to drought. *Development and Change* **9**(1): 71–86.

Birks, J.S. (1985) Traditional and modern patterns of circulation of pastoral nomads: the Duru of South-East Arabia. In: Mansell Prothero, R. and M. Chapman (eds), *Circulation in Third World Countries*. Boston, MA, Routledge and Kegan Paul.

Bishop, N.H. (1989) From Zomo to Yak: change in a Sherpa village. *Human Ecology* **17**(2): 177–204.

Bonfiglioli, A.M., R. François and M. Gomes (1988). *Nomades Peuls*. Paris, Editions Harmattan.

Bradburd, D.A. (1990) *Ambiguous Relations: Kin, Class, and Conflict among Komachi Pastoralists*. Washington, DC, Smithsonian Institution Press.

Broch-Due, V. (1983) *Women at the Backstage of Development—The Negative Impact on Project Realization by Neglecting the Crucial Roles of Turkana Women as Producer and Providers: a Socio-Anthropological Case Study from Katilu Irrigation Scheme, Turkana. (Irrigation in Arid Zones—Kenya.)* Rome, FAO.

Broch Due, V. (1990) 'Livestock speak louder than sweet words': changing property and gender relations among the Turkana. In: Baxter, P.T.W. (ed.), *Property, Poverty and People: Changing Rights in Property and Problems of Pastoral Development*. Manchester, University of Manchester, Department of Social Anthropology and International Development Centre, pp. 147–63.

Brown, B.B. (1983) The impact of male labour migration on women in Botswana. *African Affairs* **82**(328): 367–88.

Cernea, M.M. (ed.) (1991) *Putting People First: Sociological Variables in Rural Development*. 2nd edition, revised and expanded. New York, Oxford University Press for the World Bank.

Chatty, D. (1978) Changing sex roles in Bedouin society in Syria and Lebanon. In: Beck, L. and N. Keddie (eds), *Women in the Muslim World*. Cambridge, Harvard University Press, pp. 399–415.

Chatty, D. (1990) Tradition and change among the pastoral Harasiis in Oman. In: Salem-Murdock, M. and M.M. Horowitz (eds), *Anthropology and Development in North Africa and the Middle East*. Boulder, CO, Westview Press, pp. 336–49.

Cole, D.P. (1975) *Nomads of the Nomads: The al-Murrah Bedouin of the Empty Quarter*. Chicago, IL, Aldine Publishing Co.

Cole, D.P. (1981) Bedouin and social change in Saudi Arabia. *Journal of Asian and African Studies* **16** (1–2): 128–49.

Collett, D. (1987) Pastoralists and wildlife: image and reality in Kenya Maasailand. In: Anderson, D. and R. Grove (eds), *Conservation in Africa: People, Policies and Practice*. Cambridge, Cambridge University Press, pp. 129–48.

Cossins, N.J. (1985) The productivity and potential of pastoral systems. *International Livestock Centre for Africa Bulletin* **21**: 10–15.

Crotty, R.D. (1980) *Cattle, Economics and Development*. Slough, Commonwealth Agriculture Bureau.

Dahl, G. (1979) *Suffering Grass: Subsistence and Society of Waso Borana*. Stockholm, University of Stockholm, Department of Social Anthropology. (Stockholm Studies in Social Anthropology.)

Dahl, G. and A. Hjort (1979) *Pastoral Change and the Role of Drought*. Stockholm, Swedish Agency for Research Cooperation with Developing Countries.

Darling, F.F. and M.A. Farvar (1972) Ecological consequences of sedentarization of nomads. In: Farvar, M.T. and J.P. Milton (eds), *The Careless Technology: Ecology and International Development*. Garden City, The Natural History Press, pp. 671–82.

Di Domenico, C.M., and M.B. Vabi (1988) Women and dairying in Nigeria: an alternative development strategy. Paper presented at the DAWN Regional Meeting at the Institute of African Studies, University of Ibadan, Nigeria, 27–29 September.

Dyson-Hudson, N. (1991) Pastoral production systems and livestock development projects: an East African perspective. In: Cernea, M. (ed.), *Putting People First: Sociological Variables in Rural Development*. 2nd edition, revised and expanded. New York, Oxford University Press for the World Bank, pp. 219–56.

Eicher, C. (1985) Unpublished paper presented to a World Bank conference held at Bellagio, Italy, 25 February–1 March.

El-Bushra, J. (1986) *Programming for Women's Development in Sablaale, Somalia: Report of a Consultancy for EuroAction ACORD*. London: Agency for Cooperation and Research in Development (ACORD).

Ensminger, J.E. (1987) Economic and political differentiation among Galole Orma women. *Ethnos* **521–II**: 28–49.

Ezeomah, C. (1985) *The Work Roles of Nomadic Fulani Women: Implications for Economic and Educational Development*. Jos, Nigeria: University of Jos. (Postgraduate Open Lecture Series 2[8].)

Fauck, R., E. Bernus and B. Peyre de Fabrègues. (1983) *Mise à Jour de l'Etude de Cas sur la Désertification et Renforcement de la Stratégie Nationale en Matière de Lutte Contre la Désertification*. Paris, UNESCO/UNSO.

Galaty, J.G. (1988) Scale, politics and co-operation in organizations for East African Development. In: Attwood, D.W. and B.S. Baviskar (eds), *Who Shares? Co-operatives and Rural Development*. Delhi, Oxford University Press, pp. 282–308.

Goody, J. (ed.) (1962) *The Development Cycle in Domestic Groups*. Cambridge, Cambridge University Press for the Department of Archaeology and Anthropology.

Grandin, B.E. (1986) Land tenure, subdivision, and residential change on a Maasai group ranch. *Development Anthropology Network* **4**(2): 9–13.

Haaland, G. (1969) Economic determinants in ethnic processes. In: Barth, F. (ed.), *Ethnic Groups and Boundaries: The Social Organization of Culture Difference*. Boston, MA, Little, Brown and Company.

Helldén, U. (1990) Desertification—Facts and Myths. Report from the *SAREC/Lund International Meeting on Natural Desertification and Man-Induced Land Degradation*. Üorenäs Slott, Sweden, 5–7 December.

Herren, U.J. (1990) The commercial sale of camel milk from pastoral herds in the Mogadishu hinterland, Somalia. *ODI Pastoral Development Network Paper* 30a.

Hjort af Ornäs, A. (1989) Environment and security of dryland herders in Eastern Africa. In: Hjort af Ornäs, A. and M.A. Mohamed Salih (eds), *Ecology and Politics: Environmental Stress and Security in Africa*. Uppsala, Scandinavian Institute of African Studies, pp. 66–88.

Hjort af Ornäs, A. (1990) Town-based pastoralism in Eastern Africa. In: Baker, J. (ed.), *Small Town Africa: Studies in Rural–Urban Interaction*. Uppsala, Scandinavian Institute of African Studies, pp. 143–60.

Hogg, R. (1986) The new pastoralism: poverty and dependency in Northern Kenya. *Africa* **56**(3): 319–33.

Holden, S.J., D. Layne Coppock and M. Assefa (1990) *Pastoral Dairy Marketing and Household Wealth Implications and their Implications for Calves and Humans in Semi-Arid Ethiopia*. Unpublished ms.

Homewood, K.M. and W.A. Rodgers (1991) *Maasailand Ecology: Pastoralist Development and Wildlife Conservation in Ngorongoro, Tanzania*. Cambridge, Cambridge University Press.

Hooglund, M. (1977) *Report on the Eastern Senegal Bajkel Range Livestock Project*. Binghamton, State University of New York at Binghamton, Development Anthropology Seminar.

Horowitz, M.M. (1972) Ethnic boundary maintenance among pastoralists and farmers in the Western Sudan (Niger). *Journal of Asian and African Studies* 7(1,2): 105–14.

Horowitz, M.M. (1973) Relations entre pasteurs et fermiers: Compétition et complémentarité. *Notes et Documents Voltaiques* **6**(3): 42–5.

Horowitz, M.M. (1975) Herdsman and husbandmen in Niger: values and strategies. In: Monod, T. (ed.), *Pastoralism in Tropical Africa*. Oxford: Oxford University Press for the International African Institute, pp. 387–405.

Horowitz, M.M. (1989) Victims of development. *Development Anthropology Network* 7(2): 1–8.

Horowitz, M.M. (1993) River basin development policy and women: a case study from the Senegal River Valley. Paper presented to the *Symposium on the Impact of Environmental Degradation and Poverty on Women and Children*. Geneva, 27–30 May 1991. United Nations Conference on Environment and Development. Forthcoming.

Horowitz, M.M., E. Arnould, R. Charlick, J. Eriksen, R. Faulkingham, C. Grimm, P.D. Little, T. Painter, C. Saenz, M. Salem-Murdock and M. Saunders (1983) *Niger: A Social and Institutional Profile*. Binghamton, Institute for Development Anthropology.

Horowitz, M.M. and F. Jowkar (1992) *Pastoral Women and Change in Africa, the Middle East, and Central Asia*. Binghamton, Institute for Development Anthropology.

Horowitz, M.M. and P.D. Little (1987) African pastoralism and poverty: some implications for drought and famine. In: Glantz, M.H. (ed.), *Drought and Hunger in Africa*. Cambridge, Cambridge University Press, pp. 59–82.

Horowitz, M.M. and M. Salem-Murdock (1993) River basin development policy, women, and children. A case study from the Senegal River Valley. In: Steady, F.C. (ed.) *Women and Children First: Environment, Poverty, and Sustainable Development*. Rochester, VT, Schenkman Books.

Ibrahim, F.N. (1984) *Ecological Imbalance in the Republic of the Sudan—With Reference to Desertification in Darfur*. Bayreuth, Druckhaus Bayreuth Verlagsgesellschaft mbH.

Jakubowska, L.A. (1984) The Bedouin family in Rahat: perspectives on social changes. In: Marx, E. (ed.), *Notes on the Bedouin*. Sede Boqer, Institute for Desert Research.

Joekes, S. and J. Pointing (1991) Women in pastoral societies in East and West Africa. *International Institute for Environment and Development (IIED) Dryland Networks Programme. Issues Paper* No. 28. September.

Jowkar, F., M.M. Horowitz, C. Naslund and S. Horowitz (1991) *Gender Relations of Pastoral and Agropastoral Production: A Bibliography with Annotations*. Binghamton, Institute for Development Anthropology.

Kelly, H. (1986) Uncounted labour: women as food producers in an East African pastoral community. In: Moses, Y.T. (ed.), *Proceedings—African Agricultural Development Conference: Technology, Ecology, and Society, Pomona, CA, 28 May–1 June 1985*. Pomona, California State Polytechnic University, pp. 62–66.

Kerven, C.M. (1987) Some research and development implications for pastoral dairy production in Africa. *International Livestock Centre for Africa Bulletin* **26**: 29–35.

Kettel, B. (1986) The commoditization of women in Tugen (Kenya) social organization. In: Robertson, C. and I. Berger (eds), *Women and Class in Africa*. New York, Africana Publishing Company, pp. 47–61.

Kipuri, N.N. Ole (1989) *Maasai Women in Transformation: Class and Gender in the Transformation of a Pastoral Society* [*East Africa*]. PhD dissertation, Temple University.

Lancaster, W.O. (1980) Self-help for pastoralists. *Nomadic Peoples* **6**: 8–12.

Lane, C. and J.N. Pretty (1991) Displaced pastoralists and transferred wheat technology in Africa. *IIED Sustainable Agriculture Programme Gatekeeper Series* No. 20. London, International Institute for Environment and Development.

Lewando-Hundt, G. (1984) The exercise of power by Bedouin women in the Negev. In: Marx, E. and A. Shmueli (eds), *The Changing Bedouin*. New Brunswick, NJ, Transaction Books, pp. 83–123.

Lindsay, W.K. (1987) Integrating parks and pastoralists: some lessons from Amboseli. In: Anderson, D. and R. Grove (eds), *Conservation in Africa: People, Policies and Practice*. Cambridge, Cambridge University Press, pp. 149–67.

Little, P.D. (1989) The dairy commodity system of the Kismayo Region, Somalia: rural and urban dimensions. *IDA Working Paper* No. 52.

Little, P.D. (1991) *Social Issues in Rangeland Assessments*. Unpublished manuscript prepared for the World Bank.

Little, P.D. (1993) Maidens and milk markets: the sociology of dairy marketing in Southern Somalia. In: Fratkin, E., E. Roth and K. Galvin (eds), *Research on East African Pastoralism*. Boulder, CO, Lynne Reiner.

Loutan, L. and J.-M. Lamotte (1984) Seasonal variations in nutrition among a group of nomadic pastoralists in Niger. *Lancet* 945–7.

Mackenzie, F. (1990) Gender and land rights in Murang'a District, Kenya. *Journal of Peasant Studies* **17**(4): 609–43.

Marx, E. (1984) Economic change among pastoral nomads in the Middle East. In: Marx, E. and A. Shmueli (eds), *The Changing Bedouin*. New Brunswick, NJ, Transaction Books, pp. 1–15.
Marx, E. (1986a) Bedouin labour migrants of South Sinai: cash and security economics. Paper presented at *African Studies Association Conference on Migration and the Labour Market*, Canterbury, UK, April 1986.
Marx, E. (1986b) Labour migrants with a secure base: Bedouin of South Sinai. In: Eades, J. (ed.), *Migrants, Workers, and the Social Order*. London, Tavistock, pp. 148–64. (Association of Social Anthropologists Monograph No. 26.)
Massae, E.E. (1990) Women in livestock development: Tanzanian experience. Paper presented at the *Women in Livestock Development Conference,* sponsored by Heifer Project International, Little Rock, Arkansas 20–22 May 1990.
Matampash, K. (1991) The Maasai of Kenya. In: Davis, S.H. (ed.). *Indigenous Views of Land and the Environment*. Background Paper No. 10, World Development Report 1992. Washington, The World Bank, pp. 28–42.
Merryman, N.H. (1984) *Economy and Ecological Stress: Household Strategies of Transitional Somali Pastoralists in Northern Kenya*. PhD dissertation, Northwestern University.
Michael B.J. (1990) Baggara women as market strategists. Paper presented at the *American Anthropological Association Annual Meeting*, New Orleans, LA.
Michael, B.J. (1991) The impact of international wage labour migration on Hawazma (Baggara) pastoral nomadism. *Nomadic Peoples* **28**: 56–70.
Monimart, M. (1989) Women in the fight against desertification. *IIED Dryland Programme Issues Paper* No. 12.
Morton, J. (1990) Aspects of labour in an agro-pastoral economy: the northern Beja of Sudan. *Pastoral Development Network Paper* 30b. London, Overseas Development Institute.
Pouillon, F. (1990) Sur la 'stagnation' technique chez pasteurs nomades: Les Peul du Nord Sénégal entre l'économie politique et l'histoire contemporaine. *Cahiers des Sciences Humaines* **26**(1–2): 173–92.
Prior, J.C. (1992) *Planning for Pastoral Development in the Third World: Histories, the Erigavo Case Study, and Future Directions*. Master's Dissertation in Urban and Regional Planning. Department of Geography and Planning. University of New England (Australia).
Prior, S. [Suleikha Ibrahim] (1989) *The Causes, Nature, and Magnitude of Women's Reproductive Health Problems in Northern Somalia. Why Does Maternal Mortality Remain So High?* Unpublished report for Oxfam/UK.
Quechon, M. (1985) L'instabilité matrimoniale chez les Foulbé du Diamaré. In: Barbier, J.C. (ed.), *Femmes du Cameroun: Mères pacifiques, femmes rebelles*. Paris, Karthala, pp. 299–312.
Ridder, N. de and K.T. Wagenaar (1984) A comparison between the productivity of traditional livestock systems and ranching in Eastern Botswana. *International Livestock Centre for Africa Newsletter* **3**(3): 5–7.
Rodriguez, G. (1987) The impacts of milk pricing policy in Zimbabwe. *International Livestock Centre for Africa Bulletin* **26**: 2–7.
Rupp, M. (1976) *Observations sur la situation générale des éleveurs après la sécheresse*. Washington, Agency for International Development. (Projet de Range Management et de l'Elevage.)
Salih, M.A.M. (1985) Pastoralists in town: some recent trends in pastoralism in the North West of Omdurman District. *Pastoral Development Network Paper* No. 20b.
Salzman, P.C. (1987) From nomads to dairymen: two Gujarati cases. *Nomadic Peoples* **24**: 44–53.
Sandford, S. (1983) *Management of Pastoral Development in the Third World*. Chichester, John Wiley, in association with the Overseas Development Institute, London.

Schoonmaker Freudenberger, K. (1991) Mbegué: The disingenuous destruction of a Sahelian forest. *Development Anthropology Network* 9(2): 2–12.

Shoup, J. (1985) The impact of tourism on the Bedouin of Petra. *Middle East Journal* **39**(2): 277–91.

Sikana, P.M. and C.K. Kerven (1991) The impact of commercialisation on the role of labour in African pastoral societies. *Pastoral Development Network Paper* 31c. London, Overseas Development Institute.

Sörensen, A. and D. von Bülow (1990) *Gender and Contract Farming in Kericho, Kenya*. Project Paper 90.4. Copenhagen, Centre for Development Research.

Sperling, L. (1985) Recruitment of labour among Samburu herders. *East African Pastoral Systems Project Discussion Paper* No. 2. Montreal, McGill University, Department of Anthropology.

Sperling, L. (1987) Wage employment among Samburu pastoralists of North-central Kenya. In: Isaac, B.L. (ed.), *Research in Economic Anthropology*. Greenwich, CT, JAI Press, pp. 167–90.

Talle, A. (1988) Women at a loss: changes in Maasai pastoralism and their effects on gender relations. *Stockholm Studies in Social Anthropology* No. 19. Department of Social Anthropology. Stockholm, University of Stockholm.

Talle, A. (1990) Ways of milk and meat among the Maasai: gender identity and food resources in a pastoral economy. In: Pálsson, G. (ed.), *From Water to World-Making: African Models and Arid Lands*. Uppsala, Scandinavian Institute of African Studies, pp. 73–92.

Teitelbaum, J.M. (1980) *Nutrition Impacts of Livestock Development Schemes among Pastoral Peoples*. Washington, DC, USAID.

Turton, D. (1987) The Mursi and National Park development in the Lower Omo Valley. In: Anderson, D. and R. Grove (eds), *Conservation in Africa: People, Policies and Practice*. Cambridge, Cambridge University Press, pp. 169–86.

Walshe, M.J., J. Grindle, A. Nell and M. Bachmann (1991) *Dairy Development in Sub-Saharan Africa: A Study of Issues and Options*. World Bank Technical Paper Number 135, Africa Technical Department Series. Washington, DC, The World Bank.

Warren, A. (1990) Desertification as a scientific and as a social construct. Paper presented to the *International Meeting on Natural Desertification and Man-Induced Land Degradation*, Üorenäs Slott, Sweden, 5–7 December.

Waters-Bayer, A. (1985) Modernising milk production in Nigeria: who benefits? *Ceres* **19**(5): 34–9.

Waters-Bayer, A. (1988) *Dairying by Settled Fulani Agropastoralists in Central Nigeria: The Role of Women and Implications for Dairy Development*. Kiel, Wissenschaftsverlag Van Kiel. (Farming Systems and Resource Economics in the Tropics, Volume 4.)

Watson, C., M. Ezra, A. Lobuin and E. Ekuwam (1988) *The Development Needs of Turkana Women*. (First Draft.) Oxford, OXFAM/Public Law Institute.

Watson, C., M. Ezra, A., Lobuin and E. Ekuwam (1989) *Turkana Women: Their Contribution in a Pastoralist Society*. Unpublished manuscript.

Zuppan, M.E. (1992) *The Political Ecology of Agropastoralism in Burkina Faso: The Fulbe Gurmaabe of Tapoa Province*. Master's thesis in Anthropology. Binghamton University.

14 Rural Middle Eastern Women and Changing Paradigms

FOROUZ JOWKAR
Institute for Development Anthropology, Binghamton, USA

INTRODUCTION

This chapter briefly delineates the major approaches in the larger social science literature used to analyse women's roles in the Middle East, and shows how these analyses may increasingly be incongruent with the complex and changing conditions of women's socio-economic lives. It argues that although in most urban areas of the Middle East women's economic participation may still lag behind other regions, the picture is rapidly changing in rural areas where an increasing number of women are drawn into the agricultural labour force as men migrate out. Women's assumption of new roles in agrarian sectors will have environmental consequences. Additional quantitative data on women's agricultural participation are needed to substantiate the qualitative anthropological information in order to encourage the planning and implementation of environmentally sound and sustainable development actions.

DISCUSSION

Among the targets of criticism directed at the early crescendo of academic writings on women's lives has been the widespread assumption that women, as women, can be treated as a uniform analytical category. Such concepts as 'the position of women', 'the role of women', or 'the subordination of women' in a society have overgeneralized without paying adequate attention to the diverse range of activities and subjective life experiences of particular women with particular socio-economic and cultural attributes. Probably nowhere in the world has the use of over-generalizing conceptual categories gained such popularity as in the Middle East. Founded on a history of adversity between the world of Muslim Orient and that of Christian Occident, the early literature on Middle Eastern women, almost exclusively written by male travellers, drew on a particular interpretation of Islam as a libidinous religion that reduced women to sexual objects at the services of men (see Said, 1978, for a discussion of the Western image of the Middle East).

Social Aspects of Sustainable Dryland Management. Edited by Daniel Stiles

Later, the sterile dualism of traditional/modern, popular terms used to distinguish the West against the others was vigorously adopted to define the Middle East as a culturally distinct region with its own inner rules of rationality alien to Western logic and practices. Although the differences between the Middle East and the West are indisputable, the mutual exclusivity of what traditional/modern entail masked the nuances of experience that the diversity of peoples in both regions have, and substituted simplistic description for inquiries into the complexities of social orders.

In harmony with the conceptual dualism of traditional/modern, such theoretical paradigms as public and private domains and honour and shame were adopted—especially in anthropology—to deal with gender hierarchy in the regions. Women's exclusion from the male-dominated public world of politics and economics was interpreted in part to have emerged out of a context of resource scarcity and weak state power, prompting competition among different patrilineal groups who, by laying a high premium on female chastity, ensured the legitimacy of their offspring. Control of female reproductive capacity, prescribed through codes of sexual morality and formalized in religious dictums, was seen as a mechanism for maintaining group boundaries and access to resources (Schneider, 1971). The apparent physical segregation of men and women, buttressed by Islamic moral ideologies regarding each sex's activities, was often regarded as an expression of low female status in the society (Beck and Keddie, 1978: 18; Dwyer, 1987: 4–5; Youssef, 1974).

The literature about the domestic lives of women in the Middle East described how women drew power from their physical exclusion from the public life of men. In spite of references to the complementarity of the gender division of tasks in rural and pastoralist societies, however, by subsuming a wide range of women's productive activities under reproductive and domestic practices, the public/private model failed to account for the labour contribution of women in such undocumented areas as petty commodity production and commerce, domestic services, seasonal agricultural labour on family farms, or herding enterprises. By focusing on the domestic lives of women, the literature documented women's economic participation as much lower than in other parts of the developing world (see Boserup, 1970 on Muslim women; Youssef, 1974). While it linked political and ecological imperatives to a gender division of morality and labour, the theoretical focus on honour and shame ideology and public/private domains raised few questions as to how the gender division of labour is subject to change as a result of larger economic and political processes.

Influenced by debates on the effects of economic change on gender relations in developing countries, researchers interested in the Middle East have increasingly focused on the diverse range of productive activities involving women. Inspired by questions raised about the separation of concepts of production and reproduction, especially where women's domestic activities are commoditized or made invisible in the statistics, greater emphasis has been

Photograph 14.1. Control of female reproductive capacity was seen as a mechanism for maintaining group boundaries and access to resources. (Photo: Daniel Stiles)

placed on women's productive roles in the changing economies of the Middle East (Fernea, 1987). Going beyond the static categories of public/private domains, a new genre of Middle Eastern and Western researchers, while conferring importance on the role of ideology, focused on the historical construction of the gender division of labour in the region (Glavanis and Glavanis, 1989; Shami *et al.*, 1990; Morsy, 1991; Afshar, 1985; Moghadam, 1992). By circumventing such homogenizing terms as 'the status of women,' new data are disaggregated along the divisive lines of class and ethnicity, rendering information on the lives of women in the Middle East more accurate and less susceptible to stereotypes. The new researchers highlight the role of the state and the ideologies of nationalism and Muslim fundamentalism in formulating social change that politicizes and changes women's roles (Hijab, 1988; Rassam, 1984; Afshar, 1992).

What emerges out of this evolving literature is that countries of the Middle East have different economic and political formations and that their incorporations into world capitalism are uneven because of their diverse natural resources, strategic locations, and particular histories. Changes in the region have not affected all women uniformly; their integration into the process of development is a function of their class position, education, and a host of other cultural and regional factors (Mernissi, 1978; Berik, 1987; Abadan-Unat, 1981). The issue of women's integration into the process of development is by itself a thorny topic (Fuentes and Ehrenreich, 1983; Sen and Grown, 1987).

A quick glance at limited data from the Middle East illustrates fast-paced changes in the rate and patterns of women's economic participation. The most striking of these changes is the increase of working age women in the formal labour force. In the mid-1970s, the percentage of economically active women in the Muslim Middle East was less than half that in non-Muslim countries. Although the gap narrowed in the 1980s, the female labour force in the Middle East remained lower than in other areas (Mujahid, 1985; ILO-INSTRAW, 1985).

With the expansion of investment resulting from the rise of oil prices, women in the Middle East formed a larger proportion of the urban labour force, occupying positions in the service and industrial sectors, although simultaneously with a general increase in urban unemployment and underemployment (Moghadam, 1992: 94). Because the more industrialized nations of the Middle East—such as Iran, Turkey, Egypt, and Algeria—relied on import-substitution industrialization, the bulk of women's industrial employment was in low-technology cottage industries, leaving the heavier manufacturing to the male industrial labour force (Moghadam, 1988). Except in Israel, low representation of women in sales and service work may reflect the cultural sanctions not favouring close contact between females and unrelated males and the long-standing domination of men in traditional urban markets (Mujahid, 1985). Women's involvement in the paid labour market is influenced by their education, making upper and middle class more prominent in the upwardly mobile professions in administration, health, and welfare jobs that are mostly administered by the state. Lower class women's involvement is a function of their education, concentrating them in low paid and frequently undocumented sectors of agriculture, service, and urban employment (Moghadam, 1992).

Advances in information on women's economic participation in urban areas of the Middle East have been followed, however less intensively, by studies on gender relations of production in agriculture, on how gender is implicated in changes in environmental management, on links between contemporary resource use and the marginalization of women, and on the involvement of women in social movements organized to solve rural environmental problems.

Table 14.1. Female percentage of the total agricultural labour force in selected countries of the Near East Region, 1950–85

	1960	1970	1980	1985
Iran	5.0	9.4	41.0	41.0
Iraq	2.0	5.3	41.0	41.0
Jordan	3.9	4.0	0.9	1.1
Lebanon	17.5	22.7	32.2	37.7
Libya	1.7	7.2	16.0	17.7
Morocco	2.8	10.4	14.1	16.0
Oman	3.1	2.9	2.9	2.6
Saudi Arabia	3.8	3.8	3.2	3.2
Sudan	23.3	23.3	24.1	26.4
Syria	9.2	13.4	27.4	27.5
Tunisia	1.5	5.5	19.9	20.8
Turkey	49.9	48.9	51.6	54.3
Yemen AR	4.6	5.5	0.4	10.2
Yemen PDR	11.2	13.9	13.0	13.6

Source: FAO (1986).

Despite the rarity of macroeconomic data, an FAO report of fourteen different countries estimates a varied and increasing range of women's agricultural participation in the Middle East. The documented trend in macroeconomic data on women's participation in agriculture may not indicate the full range of female farm labour, in part because some countries—such as Pakistan, the Islamic Republic of Iran, and the Syrian Arab Republic—do not disaggregate their data by sex, or count their female agricultural labour force at all (Table 14.1).

Neither anthropologists nor farming systems researchers have provided as careful analyses of women's agricultural and pastoral labour in the Middle East as they have in Sub-Saharan Africa, South and South-east Asia, and Latin America. Because available information is more qualitative than quantitative (FAO, 1984: 29; Adra, 1983: 15; Myntti, 1979: 53; Tavakolian, 1986; Beck, 1978), development planning in the Middle East continues to be heavily informed by anecdotal and stereotypic information about gender relations of production. Even where anthropologists have been more directly involved in such planning, their recommendations tend either to focus on improvements in the domestic lives of women that do not confront claimed cultural values, such as providing opportunities for women in market sewing and handicrafts (Chatty, 1984), or to perpetuate gender-based economic inequalities by de-emphasizing women's access to the means of production (Salem-Murdock, 1990: 124).

Biased underestimations of women's contributions to rural production systems are a feature not only of Western models of explanation but also of Middle Eastern officials' interpretation of what women do and own, even though government personnel may informally acknowledge the importance of women's work. Authoritative denial of women's farm labour participation contributes to a paucity of policies and measures needed to improve women's productive skills and opportunities. For example, in the Islamic Republic of Iran, government officials working with a World Bank range rehabilitation project involving nomads frequently mention the dire living conditions of pastoralist women, their arduous daily labour, hardship of trek, dismal hygiene, and rampant illiteracy. But the government's primary interest is in providing training and services for men, even though women, by virtue of their involvement in milking and dairying, contribute significantly to herd and community welfare. Because of an assumption that men are the primary providers for their households, women's income-generating potential in dairying is neglected, despite the project's subsidiary goal of increasing dairy output. Instead, much of the planned training for women will be in carpet weaving, a trade infested with male middlemen exploiting vulnerable weavers. Co-operatives promoting fair wages for women's weaving activities are very few, and those in operation do not reach pastoral women.

Following the World Conference on Agrarian Reform and Rural Development (WCARRD), FAO supported ten national studies seeking information on female labour participation in rural economies of the Middle East. Results showed that women, in addition to processing and storing food, participate in the subsistence needs of their families in cropping, harvesting, sacking, thinning, planting, replanting, irrigating, manuring and fertilizing, furrowing, spraying insecticides, axe-weeding, levelling, and even ploughing (see Delancy and Elwy, 1989). Pronounced inter- and intra-regional variations in the female agricultural labour force and types of activities may be a function of seasonality, farm size, male labour migration, and cropping patterns.

Quantified results of studies similar to the FAO's are essential to substantiate the qualitative assertion of the new genre of anthropological studies describing male and female labour complementarity in reproduction of daily lives in rural and pastoral economies of the Middle East. Although the notion of gender complementarity in the recent literature on women in the Middle East has been awakening, it has not sufficiently alerted foreign and indigenous male development practitioners to systematically advance the involvement of women in development action. To be effective in policy making, further information on the economic values of women's contribution to agriculture, much of which is in the form of unpaid family labour, is essential. Focusing on indicators for measuring time use and productivity is important to shed stereotypes about the absence of women from agricultural production and management of resources in the Middle East.

Low and even negative rates of economic growth in the Middle East have resulted from declining oil revenues, rising national debts, protracted wars, the rise of an Islamic fundamentalism that discourages both domestic and international investment, and rapid population growth linked to women's low scholarization and participation in the formal labour market. Governments have had to depend increasingly on external financial and technical assistance. In

Photograph 14.2. Women's contribution to agriculture and the household economy, much of which is in the form of unpaid labour, is essential. (Photo: UNEP)

the agrarian sector, occurrences of environmental degradation contribute to declining land productivity. Despite the pressing need to feed rapidly growing and demanding populations, many governments are in fact adopting policies to increase agricultural exports in order to reduce national debts. Achievement of a satisfactory increase in per capita food production has proven elusive, however. After modest improvements in the 1970s, due in part to expanding the amount of farmland and in part to the application of new technologies, per capita food production in the region showed a 0.6% decline between 1980 and 1984, as new land was no longer available and the costs of new technologies began to exceed their returns (Delancy and Elwy, 1989).

Most states in the Middle East are new and post-colonial, and their power has been consolidated through such processes as repression, control of family life, and the imposition of measures that formalize traditional and often complex relations of property and access to natural resources. Examples of such policies are the state-orchestrated destruction of communal land use, namely *hema*, in Syria, and the nationalization of the rangelands in Iran under the Reza Shah. In the 1970s the Syrian government restored the *hema* system (Draz, 1978).

Nationalization or privatization of communal rangelands is one of the ways that states strengthen their grip on their populations, for it facilitates tax collection from and political control of the users, as it changes the agrarian systems from subsistence to commoditized production. Formal restriction on the use of the rangelands is usually accompanied by a number of changes such as reduction of areas available for herding, intensification of animal production, and expansion of farming, all in part resulting from population growth. These processes have both a direct and an indirect bearing on the gender division of labour and access to resources.

In Morocco, while herding and farming were components of a single production system in which pastoral lands were communally available while access to farmlands was on an individual basis, a series of changes is reducing the area available for herding and forcing the intensification of animal production. Other changes include population growth, expansion of farming, rapid privatization of lands, increased production for the market, shifts from cereals to orchards that do not facilitate grazing, technological changes, and, especially, male labour migration to urban Moroccan centres and overseas. As commercialization affects livestock production there are changes in herd composition—especially a movement away from goat husbandry—and farmers must produce more fodder crops to keep animals close to the farm (Bencherifa and Johnson, 1990). 'Increasingly this alfalfa-cattle loop is operated by the women, who characteristically remain at home while men are away. It is the women who cut the fodder, care for the cows, and conduct the milking operation' (Bencherifa and Johnson, 1990: 407).

With the decline in agricultural production and the accelerated increase in the prices of goods, services, and food, an increasing number of men in the Middle East migrate to cities and overseas in search of employment. As did the interregional migration of the 1970s and 1980s between oil-rich and oil-poor countries, male rural out-migration affects the local gender and age division of labour, with women, children, and older men carrying the burden of new responsibilities. Like other women worldwide, these women are the *de facto* heads of their households without enjoying ownership rights over property, yet their labour contribution to the agricultural sector is unprecedented.

Islamic laws recognize women's ownership and disposal rights over such resources as land, trees, water, and livestock. In part because of discriminatory inheritance laws that grant women half of what men inherit, and in part

Photograph 14.3. Male rural out-migration results in women, older men, and, as here in Morocco, children carrying the burden of new responsibilities. (Photo: Daniel Stiles)

because women's property rights may be abrogated by their male relatives (Pastner, 1978), Middle Eastern rural women have few enforceable rights over the ownership of resources. Women's access to resources, however, may be informal (Moors, 1989) and subject to interpersonal negotiation and to class status (Abu-Lughod, 1986). Where states establish legal codes and issue titles for the use of resources, women often get shortchanged.

In its attempt to counter land degradation and avoid problems associated with spontaneous settlement of herders on the fringes of towns, the Islamic Republic of Iran is launching a programme of planned nomadic settlement. Despite the informal acknowledgement of Iranian officials as to the importance of women's contributions to the economy of nomadic households, landownership on government-sponsored settlements will privilege only men. Interviews by the author with spontaneous settlers clearly indicated the increased role of settled women in agriculture, yet officials resist any legal transfer of land to women or to unmarried men. While unmarried men may migrate to cities in search of employment, women remain on the land as household dependents without rights to apply for loans or credit, even in the event of their husbands' death; male relatives of the deceased are said to be the guardians of widows and their orphans.

Changes in the gender division of labour and, therefore, in women's relation to resources are, of course, conditioned by the migrants' economic class and amount of remittances, the tenure system, the household development cycle and women's age, the flexibility of extended patricentric households, and the women's individual responses to change. Marx (1986), reporting on Bedouin of South Sinai, points out that when men migrate to work for wages, women and children are obliged to tend flocks and gardens. In the local systems of meanings and interpretation, women's new responsibilities are not considered 'work', yet men performing the same functions are regarded as working. Michael (1991), in her research on the impact of male out-migration on Baggara women of the Sudan, describes how because of the economic autonomy of Baggara women, who draw cash from dairy marketing activities, the income generated by labour migrants is used to expand their livestock holdings or to purchase gifts for a wide network of relatives who will act as a security buffer in case of environmental calamities.

Male out-migration and changing gender relations directly impact environmental management and resource extraction. In his evaluation of development-related demographic studies of the Sultanate of Oman, Birks states that the consequences of male labour migration include sedentarization with increases in agricultural activities based on water extracted by pumps. While women are taking a wider and more active role in labour and decision making in small gardening, a decline in the number of camels and a relative increase in the number of goats and sheep is made possible by the feminization of pastoralism. Male out-migration has also resulted in a decline in the viability of both the pastoral and agricultural sectors and the destruction, because of labour shortage and neglect, of capital investment in irrigation systems (Birks, 1976, 1985).

In Yemen, where rainfed agriculture is practised, women's agricultural activities have increased, adding a burden to their reproductive responsibilities. Forced to plough and repair terraces, women cultivate fewer fields and raise fewer livestock. On the other hand, with tubewell irrigation, lands formerly unsuited for intensive agriculture have come under cultivation of vegetables traditionally planted by women (Adra, 1983). In the past, the availability of labour ensured resource productivity. Despite the drawing of women in the absence of men into new areas of productive activities, there may be a drop in productivity. Women are rarely eligible for credit, loans, and other state-funded inputs (such as fertilizer and insecticides) necessary for sustained productivity. Access to such other services as extension and training in agricultural production is also limited for women, either because training in agriculture offered by governments and development agencies is male exclusive or because few female agriculture extension agents work in remote areas. The subjects most available in extension training are sewing and handicrafts that can be performed by women in the houses, which reinforces the stereotypes about women's separation from the public domain of production.

Because of its discriminatory policies against women's education in agriculture, the Islamic Republic of Iran is now facing a shortage of female extension staff for implementing its joint project with the World Bank on natural resource management that simultaneously attempts to rehabilitate the rangelands, improve herd productivity, and increase herders' income. With pastoral sedentarization and the growing involvement of herders in farming, it is unlikely that the goal of increasing household income will be possible to achieve if the growing role of women in agriculture is not addressed. Because settled herders are obliged to increase their herd offtake to counter environmental degradation, women will also lose out on their traditional access to wool from their family herd.

CONCLUSION

Although the social science literature on Middle Eastern women may be rich descriptively, its dominant paradigms fall short of explaining the complexity of socio-economic changes in gender relations. Such processes as integration into the global market, wars, and politicization of ideologies of nationalism and Islamic fundamentalism bring about variations in the extent of women's participation, but without doubt Middle Eastern women are becoming important players in the labour force, especially in rural areas. By assuming an expanded range of responsibilities, women are emerging as environmental managers, frequently without the *de jure* property rights that legally qualify them for access to the resources necessary for increased agricultural productivity. As the goal of feeding the growing populations in the Middle East can only be reached by environmentally sound increases in land productivity, governments and planners can no longer afford to plan and implement agricultural policies based on stereotypes. For the fragile and precious natural resources in the Middle East to be productively sustainable, women's labour in the agrarian sector must be acknowledged, and serious policy actions must be taken to enhance their legal and actual access to factors of production and channels of marketing and distribution.

REFERENCES

Abadan-Unat, N. (1981) Introduction: labour force participation. In: Abadan-Unat, N. (ed.), *Women in Turkish Society*. Leiden, E.J. Brill.

Abu-Lughod, L. (1986) *Veiled Sentiments: Honour and Poetry in a Bedouin Society*. Berkeley, CA, University of California Press.

Adra, N. (1983) The impact of male migration on women's role in agricultural production in Yemen Arab Republic. Prepared for the *Inter-Country Experts Meeting on Women in Food Production*. Amman, Jordan, 22–26 October. Rome, FAO.

Afshar, H. (1985) *Women, Work and Ideology in the Third World*. London, Tavistock.

Afshar, H. (1992) Women and work: ideology not adjustment at work in Iran. In: Afshar, H. and C. Dennis (eds), *Women and Adjustment Policies in the Third World*. New York, St Martin's Press.

Beck, L. (1978) Women among the Qashqa'i Nomadic pastoralists in Iran. In: Beck, L. and N. Keddie (eds), *Women in the Muslim World*. Cambridge, MA, Harvard University Press, pp. 351–73.

Beck, L. and N. Keddie (1978) *Women in the Muslim World*. Cambridge, MA, Harvard University Press.

Bencherifa, A. and D.L. Johnson (1990) Adaptation and intensification in the pastoral systems of Morocco. In: Galaty, J.G. and D.L. Johnson (eds), *The World of Pastoralism*. New York, Guilford Press, pp. 394–416.

Berik, G. (1987) *Women Carpet Weavers in Rural Turkey: Patterns of Employment, Earning, and Status*. (Women, Work, Development Series No. 15.) Geneva, ILO.

Birks, J.S. (1976) Some aspects of demography related to development in the Middle East with special reference to the Sultanate of Oman. *Bulletin of the British Society for Middle Eastern Studies* 3(2): 79–88.

Birks, J.S. (1985) Traditional and modern patterns of circulation of pastoral nomads: the Duru of South-East Arabia. In: Mansell Prothero, R. and M. Chapman (eds), *Circulation in Third World Countries*. Boston, MA, Routledge and Kegan Paul.

Boserup, E. (1970) *Women's Role in Economic Development*. New York, St Martin's Press.

Chatty, D. (1984) *Women's Component in Pastoral Community Assistance and Development: A Study of the Needs and Problems of the Harasiis Population, Oman. Project Findings and Recommendations*. New York, United Nations.

Collins, J. (1991) Women and the environment: social reproduction and sustainable development. In: Gallin, R.S. and A. Fergusen (eds), *The Women and International Development Annual*, Vol. 2. Boulder, CO, Westview Press.

Delancy, V. and E. Elwy (1989) *Rural Women and the Changing Socio-Economic Conditions in the Near East*. Rome, FAO.

Draz, O. (1978) Revival of the hema system of range reserve as a basis for the Syrian range development program. In: Hyder, D.N. (ed.), *Proceedings of the First International Rangeland Congress*. Denver, CO, Society for Range Management, pp. 100–103.

Dwyer, D. (1987) *Images and Self Images: Male and Female in Morocco*. New York, Columbia University Press.

FAO (1984) *Women in Food Production and Food Security in Africa*. Report of the Government Consultation Held in Harare, Zimbabwe, 10–13 July.

FAO (1986) World-wide estimates and projections of the agricultural and non-agricultural population segments, 1950–2025. Statistical Analysis Service, Statistical Division, Economic and Social Policy Division. Rome, FAO.

Fernea, E. (1987) Presidential address, 1986. *Middle East Studies Association Bulletin* **21**(1): 1–7.

Fuentes, A. and B. Ehrenreich (1983) *Women in the Global Factory*. Boston, MA, South End Press.

Glavanis, K. and P. Glavanis (eds) (1989) *The Rural Middle East: Peasant Lives and Modes of Production*. London, Birzeit University/Zed Books.

Hijab, N. (1988) *Womenpower: The Arab Debate on Women at Work*. Cambridge, Cambridge University Press.

ILO-INSTRAW. (1985) *Women in Economic Activities: A Global Statistical Survey 1950–2000*. Geneva and Santo Domingo, ILO/INSTRAW.

Kandiyoti, D. (1989) Women and household production: the impact of rural transformation in Turkey. In: Galvanis, K. and P. Glavanis (eds), *The Rural Middle East*. London, Birzeit University/Zed Books, pp. 181–94.

Marx, E. (1986) Labour migrants with a secure base: Bedouin of South Sinai. In: Eades, J. (ed.), *Migrants, Workers, and the Social Order*. (Association of Social Anthropologists Monograph No. 26.) London, Tavistock, pp. 148–64.

Mernissi, F. (1978) The degrading effects of capitalism on female labour. *Mediterranean People* 6.

Michael, B. (1991) The impact of international wage labour migration on Hawazma (Baggara) pastoral nomadism. *Nomadic Peoples* 28: 56–70.

Moghadam, V. (1988) Women, work, and ideology in the Islamic Republic of Iran. *International Sociology* 2(2).

Moghadam, V. (1992) Women, employment, and social change in Middle East and North Africa. In: Kahne, H. and J.Z. Giele (eds), *Women's Work and Women's Lives: The Continuing Struggle Worldwide*. Boulder, CO, Westview Press.

Moors, A. (1989) Gender hierarchy in a Palestinian village: the case of Al-Balad. In: Glavanis, K. and P. Glavanis (eds), *The Rural Middle East: Peasant Lives and Modes of Production*. London, Birzeit University/Zed Books, pp. 195–209.

Morsy, S. (1991) Women and contemporary social transformation in North Africa. In: Gallin, R.S. and A. Fergusen (eds), *Women and International Development Annual, Vol. 2*, Boulder, CO, Westview Press.

Mujahid, G.B.S. (1985) Female labour force participation in Jordan. In: Abu Nasr, J., N. Khoury and M. Azzam (eds), *Women, Employment, and Development in the Arab World*. The Hague, Mouton/ILO.

Myntti, C. (1979) *Women and Development in Yemen Arab Republic*. Germany, German Agency for Technical Cooperation.

Pastner, C.M. (1978) The status of women and property on a Baluchistan oasis in Pakistan. In: Beck, L. and N. Keddie (eds), *Women in the Muslim World*. Cambridge, MA, Harvard University Press, pp. 434–50.

Rassam, A. (1984) *Social Science Research and Women in the Arab World*. London, UNESCO and Frances Pinter.

Said, E. (1978) *Orientalism*. London, Routledge and Kegan Paul.

Salem Murdock, M. (1990) Household production, organization and differential access to resources in Central Tunisia. In: Salem-Murdock, M. and M. Horowitz, with M. Sella (eds), *Anthropology and Development in North Africa and the Middle East*. Boulder, CO, Westview Press.

Schneider, J. (1971) Of vigilance and virgins: honour, shame and access to resources in the Mediterranean societies. *Ethnology* 10: 1–24.

Sen, G. and C. Grown (1987) *Development, Crises, and Alternative Visions: Third World Women's Perspectives*. New York, Monthly Review Press.

Shami, S., L. Taminian, S.A. Morsy *et al.* (1990) *Women in Arab Society: Work Patterns and Gender Relations in Egypt, Jordan, and Sudan*. Providence, RI: Berg Publishers Ltd/UNESCO.

Tavakolian, B. (1986) Women's work and social reproduction among Sheikhanzai nomads of Western Afghanistan. Paper presented at the American Anthropological Association, 85th Annual Meeting—*Session on Gender and Social Change among Pastoralists*, Philadelphia, PA, 3–7 December 1986. (Denison University—*Global Studies Working Paper* No. 10.)

Youssef, N. (1974) *Women and Work in Developing Societies.* Population Monographs Series. Berkeley, CA, University of California Press.

15 The Impact of Social and Economic Change on Pastoral Women in East and West Africa

JUDY POINTING
Institute of Development Studies, University of Sussex, UK

INTRODUCTION

The literature on women in pastoral societies in Africa is relatively limited when compared to the vast body of research on women in agriculture in the same region. Although more attention is now being paid to the role of women in the pastoral economy, most of this literature is in the form of case studies, and different authors have tended to pursue different facets of women's position. Recently, efforts have been made to analyse this literature to develop a wider perspective on pastoral women in Africa, as well as in other parts of the world (Joekes and Pointing, 1991; Horowitz and Jowkar, 1992).

This chapter draws on a number of case studies to investigate the impact of social and economic change on pastoral women in East and West Africa. The limitations of the literature make it difficult to attempt any systematic comparison. For this reason a different approach is adopted, focusing on the similarities between the two regions. There do seem to be much stronger similarities than differences in the position of women in pastoral societies, in terms of traditional rights and entitlements and the gender division of labour.

The chapter takes as its starting point recognition that pastoral societies are not isolated entities; they interact with and are in many respects fully integrated into the national economy. The crucial contribution of women's work to the success of the pastoral enterprise is also acknowledged. As such, they are neither removed from nor immune to the wider changes which are taking place.

THE DIVISION OF LABOUR IN PASTORAL SOCIETIES

An overall division of labour between women and men is prevalent in pastoral societies, although there are variations between groups and shifts do occur over time. The prevailing view of pastoral men's and women's roles has tended

Social Aspects of Sustainable Dryland Management. Edited by Daniel Stiles

Photograph 15.1. Men, as here in Mali, are responsible for herding livestock. (Photo: Daniel Stiles)

to ignore these complexities: men's work has generally been associated with herd management and women's work with the children and house. As a result the extent of women's involvement with livestock has frequently been underestimated. In fact, the balance of work is such that women frequently spend more time than their husbands in animal care (Talle, 1988: 11).

Women are closely involved in caring for young and sick livestock as well as for animals kept near the homestead. As 'milk managers' they are responsible for milking, processing milk products and marketing of dairy produce. Women also perform all domestic chores including food preparation and collecting firewood and water. They are responsible for child rearing and usually for food provision. Nomadic women also dismantle and rebuild their houses when the herd is moved to new pastures. Men's responsibilities as herd managers include moving, feeding and watering the herds, castration, vaccination and slaughter, building enclosures and digging wells. Senior men are responsible for planning and decision making with regard to livestock, while junior men and boys perform most of the physical labour and herding.

Despite overall similarities, variations occur in different cultural and economic contexts. For example, where social convention constrains women's mobility, as among the Tuareg, it is men rather than women who go to market and make household purchases (Oxby, 1978: 284). In contrast, where this convention does not apply, as among the Maasai, women have greater freedom of movement and are more involved in the sale of milk products (Talle, 1988: 226).

As has been observed in agricultural settings, shifts in the division of labour are often associated with male consolidation of control of assets which are increasing in value. While women retain their traditional involvement in small-scale dairying, men are more likely to become involved when it becomes a larger-scale, more profitable activity (Talle, 1988: 227–8). There is a tendency for men to become involved in house construction as sedentarization encourages the building of more permanent structures (Talle, 1988: 252). This may benefit women by relieving them of the on-going and time-consuming work of house building and repair, but it also excludes them from ownership of an important asset.

For the most part, however, the direction of change has been towards women taking on formerly 'male' tasks. This is attributable partly to a shortage of household labour due to greater mobility and the involvement of men in waged employment. The decline in household size and coherence has also reinforced women's work burdens. Evidence of increased involvement of young wives in herding activities in some areas has been related to the breakdown of residence structures into smaller units and greater self-sufficiency of the family (Talle, 1988: 251). These additional labour demands are placing a heavy burden on many women.

LIVESTOCK ENTITLEMENTS

The central importance of livestock in pastoral regions derives from the high-risk nature of the environment: the mobility of animals makes them less vulnerable than crops to localized drought. As livestock are pastoralists' means of survival, they play a central role in the economic, social and cultural lives of pastoral societies. The complexity of livestock rights is demonstrated by the fact that different people have different rights vested in the same animal (Broch-Due *et al.*, 1981: 253).

Women are associated with livestock as the means of subsistence as 'milk managers', a role intimately connected to their reproductive and household provisioning roles. As milk managers women control the distribution of milk between animals and humans. This balancing of animal and human needs was the crux of a successful pastoral enterprise, and the decisions taken by women with regard to milk off-take were critical to the well-being of both (Talle, 1988: 251).

Rights to milk depend on women having a reproductive role, as child-bearing establishes a woman's claim to milk. Milk and milk products can be used in a variety of ways: for consumption within the household or for exchange and marketing. Whether marketed, exchanged or consumed at home, milk products contribute directly to household welfare. Women usually have the right to retain any cash generated by the sale of dairy produce. The

Photograph 15.2. Women are in charge of milking and usually have the right to retain any cash generated by the sale of dairy produce. (Photo: Daniel Stiles)

evidence suggests that this income is generally used to meet household requirements rather than for personal consumption. Women also utilize their dairy management roles to build up their own social networks, either by giving dairy products to other women or by allowing them the use of one of their own milking animals. In this way reciprocal links are built up to maintain the flow of food into the house during periods of stress (Kettel, 1988: 8).

Women also own animals, sometimes even large herds. These are often obtained at marriage and also through inheritance, although women usually inherit less than men. Women frequently leave their animals in the care of their brothers. As it is from brothers that women derive support after the death of their father, this can be understood as an instance of asset dispersal and strengthening of family obligations that can be drawn on in times of need (Baroin, 1980: 3). Market transactions involving livestock are usually controlled by men, although they cannot freely dispose of animals in which women or children have rights (Kettel, 1988: 10).

Although herd management is a male domain, women are often involved in discussion and decision making relating to livestock, especially if they or their children have rights to particular animals. In this way women play an important role in safeguarding their children's interests in livestock. While ownership and ability to freely dispose of livestock is circumscribed by male management of herds, their traditional rights to animals do provide women with certain assets and a degree of leverage which can be used to their advantage in both the shorter and longer term (Kettel, 1988: 9; Talle, 1988: 248).

Certain changes taking place in the pastoral economy are fundamentally altering women's rights and access to livestock. In some areas, sedentarization and degradation of grasslands means that the herds tend to be kept at cattle posts in remote areas, away from the homesteads. Furthermore, the growing importance of beef production and marketing of animals is adversely affecting women's property rights in livestock. There is an increasing tendency for men to appropriate women's rights to livestock without negotiation or permission as was traditionally required. Women now complain that their animals are among the first to be sold (Ensminger, 1984: 64). It has been argued that commercial livestock transactions have made it possible for men to redefine or disregard traditional rights accruing to women and children (Talle, 1988: 266).

The growing emphasis on commercial meat production as opposed to dairy production has particular consequences for women's household provisioning responsibilities. Little of the income from commercial meat production 'trickles down' to women even though they retain their traditional responsibilities to feed their children (Kettel, 1988: 11). At the same time, diminishing supplies of milk mean they either have to allocate less to household needs or sell—and earn—less. In either case, a reduction in the supply of milk has negative implications for household nutritional status and women's labour time.

Women have demonstrated their resourcefulness in adapting to and stretching diminishing resources. Diluting milk or adding value to milk products through processing are ways women maximize earnings from reduced quantities (Waters-Bayer, 1985: 16). However, this usually demands greater time and labour input, particularly of poorer women who are more likely to be involved in food processing as a source of cash income.

It does not necessarily follow either that a greater contribution by men to household provisioning will improve household nutritional status or reduce the labour demands on women. Women and men often have quite different priorities for cash expenditures and household needs. There is frequently conflict between spouses when expenditure on food or distribution of milk is controlled by men. Women complain about men's reluctance to pay for household requirements and their expenditure on drink, or that they do not receive sufficient milk to meet the household's needs (Tallo, 1988: 265; Waters Bayer, 1985: 7; Nestel, 1985: 198).

COMMON PROPERTY RESOURCES AND USE RIGHTS

Land use in pastoral societies was traditionally governed by a complex and sophisticated set of rules and institutions, which provided access to vital although often precarious resources. The co-operative use of resources can be understood as a means of minimizing risk in a climatically unpredictable environment: the spatial and temporal variability of precipitation and consequent vegetative cover without regard to land boundaries is a defining characteristic of pastoral arid lands. The pastoralist strategy of maximization and dispersal of assets was an adaptive mechanism to these circumstances which permitted survival in times of crisis and allowed for economic regeneration.

The use of land by pastoralists as a common property resource has come under increasing pressure. Large tracts of pastoral land were appropriated during the colonial era, while in more recent years the creation of national parks has been accompanied by restrictions on pastoralists' use of this land. Population increase and growing land pressure has also led to encroachment by the agricultural sector onto rangeland. The greatest loss of commonly managed land, however, has occurred through the adjudication of grazing land into private ranches and farms. Typically, usufruct rights based on residence are replaced by legal formalization of group or individual ownership of land.

The privatization and individualization of land is a critical change in the rural economy, and has far-reaching implications for poorer households and women in particular. It has been argued that the creation of group ranches intensifies social and economic stratification as they have an inbuilt tendency towards the creation of wealth. This is because even if equal grazing quotas are allocated to members originally, those with more starting capital are likely to lease or buy quotas from other members (Hedlund, 1979: 33).

Furthermore, group ranch membership is also entirely male. It seems that pastoral women experience discrimination similar to that faced by women farmers in Sub-Saharan Africa, where land is usually registered in the name of the male household head regardless of who actually farms the land or makes the greatest economic contribution to the household. Women are as a consequence excluded from access to loans and extension services which are usually determined by land title. In this way important productive resources become concentrated in the hands of men (Staudt, 1975/6).

Substantial changes in utilization rights to water sources are also occurring with the development of wells and water tanks to provide permanent water supplies. It is becoming common for access to water to be marketed, either for cash or in exchange for livestock. Even water for human consumption is being sold. Previously water was regarded in the same way as pasture and every household had the right to draw water for animal and domestic use. Now, women tend to accumulate water debts during the dry season which they pay off when milk—and therefore their cash income—becomes more abundant (Talle, 1988: 53). It seems that women become responsible for domestic water costs when what was previously a free resource is transformed into a transactable commodity.

ENVIRONMENTAL DEGRADATION

The multiplicity of women's tasks involves them in a close interaction with and dependence on the natural environment in a number of ways, in collecting wood and water, and foraging for both animal and human consumption. There are, therefore, particular repercussions for women—especially poor women—as a result of diminished availability of rangeland resources.

Degradation of pasture land increases the amount of time that has to be spent caring for young, sick, and feeble livestock which are kept at the homestead. It is women who are responsible for collecting water and fodder for these animals. This work is particularly time consuming in the dry season when more animals are in a weakened condition, and at the same time there is a greater scarcity of fodder (Dahl, 1979: 64). Degradation of the pasture lands contributes to the deterioration of both animals and fodder supply, a combination which considerably increases the burden of work on women.

Any deterioration in the quality of grazing is rapidly translated into a reduction in milk yields. Pastoralists are therefore extremely sensitive to differences and variations in pasture because they can monitor marginal differences by measuring milk output. Pasture degradation has a particular impact on women by further reducing the milk supply which is so critical to household provisioning.

Wood shortage is another aspect of resource depletion which has particular repercussions for women. Women are responsible for collecting firewood and often for house-building. These have become increasingly time-consuming and tiring tasks as longer distances must be walked to find and gather sufficient firewood and construction material.

The shortage of wood is exacerbated by two trends which have increased demand. First, as sedentary women tend to build larger and more numerous houses than nomadic women, more wood is required for the building and repair of their homes (Ensminger, 1984: 65). Women often have to travel great distances to find wood suitable for house-building. Second, the dietary shift from milk to grain has increased the amount of cooking fuel required by households (Ensminger, 1984: 65). Purchasing fuelwood is one way of responding to the shortage for those who can afford to. For poorer households reducing energy consumption is often the only option. Shortage of firewood can, therefore, have a major influence on the nutritional status of poorer households, as women cook fewer meals or turn to foods which are nutritionally inferior but which require less cooking time (Bradly, 1991: 220).

Photograph 15.3. Near Khartoum in Sudan, Nile reeds are used for fuel as wood is rare and expensive. (Photo: Daniel Stiles)

Land privatization and environmental degradation also result in restricted availability of wild foods such as berries, fruit, plants, and roots. Many of these resources are also important for medicinal and other purposes. Wild foods are mainly collected and eaten by women and children, and provide valuable additional nutrients to an often unvaried diet (Talle, 1988: 56; Broch-Due, 1983: 173).

Nomadic women face additional problems, as a deterioration of the rangeland necessitates more frequent moves to find new pasture. House-moving is women's responsibility, and more frequent moves means that this activity becomes much more time-consuming (Dahl, 1979: 64). Conversely, where women's time is already overstretched the household's mobility is constrained. It has been suggested that shortage of labour sometimes leads to detrimental changes in management practices. Where the demands on women's time have become very severe the mobility of the main camps and their associated herds of milk animals may decrease (Dahl, 1979: 86).

The evidence thus indicates that environmental degradation contributes significantly to women's work loads while reducing their capacity to meet their household provisioning obligations. It suggests that the extra time many women have to spend in subsistence activities such as gathering wood, water, and fodder reduces the amount of time available for other economic activities.

For poor women, or those with limited access to resources, the impact is likely to be even greater.

CONCLUSION

Analysis of the situation of women in pastoral societies in East and West Africa indicates that there is a remarkable consistency in both the social and economic transformations which are taking place, and the impact of these processes on women. There are further parallels between the situation of women in agricultural and pastoral societies.

Increased monetization of the rural economy, commercialization of production, and adjudication of common property resources into private ownership are contributing to the breakdown of traditional entitlements and obligations which formerly governed social and economic relations. Women's traditional rights have proved to be vulnerable in the face of these changing social and economic conditions. This vulnerability can be related to the fact that women's traditional rights were usufruct rights, while ultimate control of resources was invested in men. Now, changes in the pastoral economy which have intensified competition for resources have led to women becoming excluded from access to productive assets. At the same time their work burdens are becoming heavier as their own workloads increase and they take on more of the tasks previously carried out by men.

Among pastoralists as in agricultural societies, allocation of tasks and responsibilities as well as access to natural and other resources are largely determined by gender. Pastoral women carry major responsibility for household provisioning and child rearing, as well as being substantially involved in livestock care. It is their involvement with livestock which forms the basis for much of women's social and economic support. Yet as the importance of subsistence milk production declines in favour of commercialized meat production their claims on livestock and livestock products are weakened. As a consequence many women are finding it increasingly difficult to meet their household provisioning obligations. This difficulty is compounded as the status and limited authority conferred on women through their close involvement with livestock is undermined.

While there are remarkable similarities in the changes taking place in pastoral societies, the impact on individual women and men varies in different contexts. The processes of change are filtered through an existing set of social and economic relations which means that some women are better placed than others to respond to increasing demands or new opportunities.

POLICY GUIDELINES

The foregoing analysis suggests a number of areas where interventions could usefully be developed. However, it should be remembered that the analysis focused on the similarities in the processes taking place in the regions surveyed. One such process has been an increase in social and economic stratification among pastoralists. These widening social and economic differences, when added to variations in environmental, cultural, and political conditions, make application of universal solutions impossible. I would argue that to be effective interventions must be developed with the full participation of the intended beneficiaries.

Four main areas for action have been identified:

(1) Safeguarding women's access to productive resources. Access to productive resources must be protected as the system of entitlements breaks down, to prevent women being marginalized as previously common property resources become individualized assets. The extent of women's involvement in both pastoral and agricultural production is now widely recognized. If, as is commonly argued, individualized ownership increases the incentives to maintain productive resources, there are sound environmental (as well as equitable) grounds for ensuring that women are not excluded from the process. Further, concentration of land ownership in the hands of men and the exclusion of women has been identified as an impediment to investment in the land by those who work it, as discussed above.
   (a) Ensuring that women are included in the land-registration process is an immediate priority. Title should be protected through laws governing marriage, divorce and inheritance;
   (b) Women's ownership of livestock should be protected in the same way;
   (c) Extension and veterinary services should be made accessible to women through (i) training female extension workers to work specifically with women; and (ii) training village women as livestock and agriculture specialists, working on a cost-recovery basis;
   (d) Providing women with access to credit and savings facilities, taking account of women's lack of mobility, lack of ownership of assets which can be used to secure loans, and a generally good record as borrowers and savers.

(2) Reducing the amount of time women spend on domestic labour. This would reduce the time constraint many women face and thereby increase the time available for income-generating activities. However, care must be taken not to displace poor women who earn income by performing these tasks for wealthier women. It should also be recognized that domestic tasks do not necessarily represent drudgery for women—household maintenance is often an area where women have a degree of autonomy. Freeing up time for more involvement in income-generating activities may not benefit women if it means as much work plus a tightening of male control over their labour and its proceeds. Four main areas have been identified:
   (a) Provision of water points for domestic use would mean a considerable saving of women's time and energy, in addition to reducing poor households' expenditure on water and improving health standards;

(b) Easier access to fuelwood would save collection time and curtail environmental damage. The creation of village woodlots, where practicable, could also provide employment for poor women. Alternatives to woodfuel, and the use of energy-efficient stoves, should also be investigated in collaboration with the women who are to use them;

(c) Access to maize-milling facilities could result in a major reduction in time spent on this arduous activity, particularly for poor women (as poor households are more dependent on grain as a staple). This could involve small-scale individual household mills, or larger-scale fixed mills in villages which are within reach of a large number of women for most of the year. This would also have the potential for providing employment for some women, and opens up the possibility of marketing grain-based products for women with limited access to milk. Improved processing would have the additional benefit of making grain more digestible, with positive implications for the health and nutrition of children in particular;

(d) Attention could also be given to ways of reducing the amount of time nomadic women must spend dismantling, rebuilding and repairing their houses through better access to building materials or design innovations, again with the full participation of women.

(3) Improving income-generating opportunities for women by:

(a) Strengthening existing areas of involvement. Women's involvement in small-scale dairying should be encouraged and supported, as a means of providing for their families and maintaining their traditional rights to and involvement in the herds. The benefits to women of involvement in their own dairying business, as opposed to selling to commercial dairies, has been demonstrated. Assistance can be given in the form of provision of clean water, improved transport and roads, improved processing and containers. Support of women's dairying enterprises should particularly aim to enhance women's control over their own labour and the proceeds derived from it;

(b) Developing new opportunities. The need for increasing alternative income-generating opportunities for women is indicated by the growing dependence of pastoral households on non-pastoral income, and the declining availability of milk both for feeding their families and marketing.

(4) Enhancing women's involvement in decision making. Efforts should be made to enable women to retain control of their earnings which, research has indicated, are largely spent on improving the welfare of their families. This could be achieved by promoting existing areas of women's involvement and control. Women's management and marketing of milk products is a good example of the benefits to be derived from involvement in the whole production and marketing cycle. The study does, however, indicate that there is a scale problem: as such enterprises become larger and more profitable they become vulnerable to appropriation by men. Collective enterprises by groups of women might be one method of overcoming this.

Photograph 15.4. One area where pastoral women can improve their living conditions is that of income-generating activities such as basket weaving. (Photo: Daniel Stiles)

NOTE

This chapter is based on Drylands Network Programme Issues Paper No. 28 published with Susan Joekes in 1991.

REFERENCES

Baroin, C. (1980) The generalized exchange of livestock: the Tubu economic system (Chad, Central Africa). University of East Anglia, *Development Studies Discussion Paper* No. 74.

Bradley, P.N. (1991) *Woodfuel, Women and Woodlots, Volume 1: The Foundations of a Woodfuel Development Strategy for East Africa*. London, Macmillan.

Broch-Due, V. (1983) *The Fields of the Foe: Amana Emoit: Factors Constraining Agricultural Output and Farmers' Capacity for Participation*. Department of Sociology, University of Bergen.

Broch-Due, V., E. Garfield and P. Langdon (1981) *Women and Pastoral Development: Some Research Priorities for the Social Sciences*. Unpublished ms.

Dahl, G. (1979) *Suffering Grass: Subsistence and Society of Waso Borana*. Department of Anthropology, University of Stockholm.

Ensminger, J. (1984) Theoretical perspectives on pastoral women: feminist critiques. *Nomadic Peoples* **16**.

Hedlund, H. (1979) Contradictions in the peripheralisation of a pastoral society: the Maasai. *Review of African Political Economy* No. 15/16.

Horowitz, M. and F. Jowkar (1992) Pastoral women and change in Africa, the Middle East and Central Asia. *Institute for Development Anthropology*, Working Paper No. 91, Binghamton, New York.
Joekes, S. and J. Pointing (1991) Women in pastoral societies in East and West Africa. *International Institute for Environment and Development Drylands Networks Programme, Issues Paper* No. 28, September.
Kettel, B. (1988) *Women and Milk in African Herding Systems*. Mimeo, York University, Toronto.
Nestel, P. (1985) *Nutrition of Maasai Women and Children in Relation to Subsistence Food Production*. Unpublished PhD thesis, Queen Elizabeth College, University of London.
Oxby, C. (1978) *Sexual Division and Slavery in a Twareg Community: a study of Dependence*. Unpublished PhD thesis, School of Oriental and African Studies, University of London.
Oxby, C. (1986) Women and the allocation of herding labour in a pastoral society: Southern Kel Ferwan Twareg, Niger. In: Bernus, S. *et al.* (eds), *Les fils et le neveu. Jeux et enjeux de la parente touareque*. Cambridge, Cambridge University Press.
Staudt, K. (1975/6) Women farmers and inequalities in agricultural services. *Rural Africana* No. 29.
Talle, A. (1988) Women at a loss: changes in Maasai pastoralism and their effects on gender relations. University of Stockholm, *Studies in Social Anthropology* No. 19.
Waters-Bayer, A. (1985) Dairying by settled Fulani women in Central Nigeria and some implications for dairy development. Overseas Development Institute, *Pastoral Development Network*, Paper 20C.

16 A View from Within: Maasai Women Looking at Themselves

BIRGET DUDEN
Elangata Wuas Ecosystem Management Programme, Kajiado District, Kenya

INTRODUCTION

This chapter is based on experiences in a project implemented within the Elangata Wuas Ecosystem Management Programme by the local Maasai self-help group, the National Museums of Kenya and Kenya Wildlife Service. The main reason for this project—also called 'photo-appraisal'—in which Maasai women themselves are taking photographs of positive and negative aspects of their lives, was the fact that for a long time the literature created the myth of powerless pastoral women (Merker, 1910; Hollis, 1970; Arhem, 1985). Male researchers as well as male informants underestimated women's involvement with livestock, and their background influence in decision-making processes.

A possible reason for this kind of superficial analysis may be the fact that the female sphere, especially in Maasai society, is not directly connected with the public, not directly approachable from outside. The problem of approaching women is based on their special role within the society. An overall division of labour as well as of responsibilities pushes the women into a background position with regard to public affairs, and eventually creates inhibitions to discussing certain subjects with outsiders.

Most researches and investigations dealing with Maasai culture are based on the structural organization of Maasai society, regarding the hierarchical division within the age-set system as social reality, and as a main guideline to collect information (Jacobs, 1973; Evangelou, 1984). Women and girls are, however, separated from the male age-set system and are not seen as a communal group, but as individuals, and have, therefore, no official function within the political system (von Mitzlaff, 1988: 73). This may lead to the wrong conclusion that women play no role in public life. Nevertheless, there are women having a position, defined responsibilities and duties within the community, and they have their own opinions, which are worth investigating.

A process of active involvement in which attention is focused not on the individual woman but on a focal point away from the individual (the photo)

Social Aspects of Sustainable Dryland Management. Edited by Daniel Stiles

helps to overcome inhibitions, and to create a relaxed atmosphere in which fruitful discussions can take place. Even if some authors claim that women are often involved in discussions and decision making, at least regarding livestock, my personal experience shows that they are not really used to discussing community affairs within a group. The main objective is, therefore, to create a situation where women can get used to discussing topical and general problems in a group.

THE PROCESS

In several *enkangs* (settlements) in the whole area of L'oodokilani (Kajiado District) women were asked to participate in the project. In each *enkang* four women were asked to take photos with disposable cameras. At the beginning of each photo-activity a discussion took place about the issues to be tackled (positive and negative aspects in the life of Maasai women) and about the possible realization of the relevant subjects. After the women identified the subjects and the way of representing them, they then took the photos themselves.

After the photos have been processed another discussion takes place, this time about the motivation to take a certain photo and about the subject touched. One photograph at a time is shown to the same group of four women. The women should identify the represented problem and talk about their attitudes and their feelings regarding this problem.

At a certain point, when four or five groups of women have already taken part in the project, a workshop is organized, in which the women discuss the presented topics with each other. Individual opinions on single aspects are given, and potential solutions for specific problems within the group are discussed.

FEMALE PERSPECTIVES

To gain a certain knowledge about culturally determined attitudes, and interactive processes between individuals and the environment, it is necessary to recognize the cultural variability of personality and, therefore, of perception patterns. The traditional way of 'seeing things' and the knowledge concerning certain environmental aspects can be expressed very clearly by taking photographs and talking about their realization and meanings.

The Maasai way of perception reflects the characteristic elements of their cultural and personality patterns and, therefore, offers the possibility to obtain information which not only includes the intellectual process of interpretation but also gives a visual impression of the 'cultural points of view'. Issues, of course, vary with the given outside situation. Main issues presented in the photographs are the time-consuming and tiring female tasks: fetching water, collecting firewood, constructing houses, care for livestock.

Previously, water was regarded as communal property. Every household had the same right to draw water for domestic and animal use. But what was previously a free resource has now in many cases become privately owned property. Women not only have to walk far to get water for domestic use, especially in dry seasons, but obviously become responsible for the water costs, which they try to pay for through their 'milk-income'. Since any deterioration in the grazing quality is directly measurable in the reduction of milk yields, water debts are accumulated during dry periods.

The availability of wood is another problem in household management. Women are responsible for collecting firewood and house-building and therefore any shortage of wood due to degradation has particular disadvantages for women.

The above-mentioned female tasks are regarded as a problem mainly during dry seasons. During rainy seasons water is available from small dams in the neighbourhood of the *enkang*. No long walks are necessary. House construction is carried out not in these periods but mainly during dry periods. During the wet seasons women are permanently busy with smearing and plastering leaking roofs. Many Maasai women would prefer to live in concrete houses rather than in the *enkaji* (traditional house). Such houses would release them from perpetual mending.

The building of modern houses even affects the traditional division of labour. Responsibility for modern house building is gradually being taken over by men. However, Kenya Maasai men have not begun to do physically demanding work of collecting and putting up the poles which support the house. Instead they employ workmen, and still sometimes rely on their wives. Obviously, rights of property of a modern house are in the hands of men. This means a change in the position of the house as the centre of the female domain.

Wet seasons raise other kinds of problems to deal with. An important task is represented by the struggle against ticks. Not much governmental support is available to purchase the necessary veterinary medicine to prevent tick fever and other diseases. Since care for livestock falls under the female responsibilities, the women have to remove the ticks by hand.

Cash income could resolve some of these problems, but the women complain that they have been increasingly excluded from control over family resources, and, therefore, have fewer possibilities to earn money by selling livestock products. Because of the wide range of their responsibilities, and their close connection to the natural environment, women have very concrete ideas about environmental management.

Among women there is a noticeable opposition to sub-division and privatization of communal land. Maasai men become influenced by outsiders, who lead them to believe in the positive aspects of sub-division by showing them photos of green pasture and fat cows. Women clearly define the borders of reason and feasibility (Talle, 1988: 52ff.).

Sub-division would leave many families with pieces of land not large enough to sustain their herds. Movements of herds would be stopped. Land degradation would increase due to the concentration of cattle in certain areas, and land would be sold to outsiders. The only positive aspect mentioned by the women would be the increase of male responsibilities to take care of the land. Herding cattle on community land demands few responsibilities in land management, except to change pasture when grazing is exhausted. The possibility of moving on with their cattle releases men from trying to prevent overgrazing and soil erosion, which they would have to do if the land was their private property.

Female awareness of these problems seems to be much higher than that of men, but the multiplicity of women's duties does not allow them to spend much time on the issue of land degradation. It can be established that land privatization and environmental degradation contribute already significantly to women's work load. Efforts should be made to allow them to break this vicious circle.

Another positive issue which is often mentioned with some concern is the cultural sphere. Being 'Maasai', keeping the traditions, and performing rituals and ceremonies is seen as positive. Maasai ceremonies are said to have always been celebrated in the same way, and they are reminders and remainders of the old times. Many women appear to be concerned because Maasai men who lose their economic and political base often adopt habits and norms which differ from those valid in the pastoral context. Lack of recognition and destitution are factors contributing to the increasing drift of Maasai men to towns and trading centres. The 'town boys' are very often regarded as 'lost Maasai'—they are not proper Maasai in the sense of being 'cattle people'.

GENDER RELATIONS

General notions like 'the role of women' have often been used to describe the female part of a given society without paying adequate attention to actual life experiences of individual women and their socio-economic attributes. There are two particular statements which characterize the different views of the 'female role' within Maasai society quite well.

On the one hand, there is the statement of the German captain Merker, emphasizing the subordination of Maasai women:

> Fuer den eigenen Mann ist die Frau besonders wichtige Arbeitskraft, wolche er fuer Haushalt und Viehwirtschaft braucht, und das Mittel zur Erfuellung seines Wunsches nach einer moeglichst grossen Nachkommenschaft. Danach, wie sie diesen Aufgaben gerecht wird, richtet sich ihre Behandlung. Maewssige Pruegel sind nicht seiten, rohe Behandlung kommt dagegen fast nie vor . . . Im oeffentlichen Leben steht die Frau, ebenso wie ihrem eigenen Mann gegenueber rechtlos dar' (Merker, 1910: 120).[1]

On the other, there is the more complex image of the ethnologist Fosbrooke:

> It is important for a number of reasons to know who holds the purse-strings in a Maasai family group, but it is a question to which no definite answer can be given. Rules, where they exist, are seldom followed—in such intimate family matters it is personality that counts (Fosbrooke, 1948: 47).

This area of tension between patriarchal structure and daily life experience has to be taken into account, with reference to the actual gender relations. As Merker pointed out, gender relations are ideally characterized by control (*aitore*) of men over women's labour, sexuality, and fertility, with the consequence that women have limited rights with regard to their own persons and bodies. Maasai society defines only two informal age-categories of women, the uncircumcized girls (*entito*), and the circumcized (married) women (*enkitok*); whereas men are organized in three formal age-sets defining the hierarchical structure of the whole society, with the elders as holders of responsibility and power (von Mitzlaff, 1988: 75).

Each male age-set is subject to certain behavioural rules which are extended to the female part of the community. Women have to follow the rules that govern male consumption of food and male sexuality, being able to gain social status only by becoming a mother (Arhem, 1985: 11).

Apart from this ideal organization of Maasai society, women play an important role in the community concerning labour, trade, child rearing, etc. and have as individual persons a good insight into communal affairs and environmental problems. Often outsiders tend to divide Maasai society into a male domain, which concerns mainly livestock management, and a female domain, concentrated on the house, limiting their own view on reality (Evangelou, 1984; Arhem, 1985).

The findings within the described project have shown that women have many duties connected with livestock, e.g. care for sick animals, herding of small stock, removal of ticks, etc. They also have special rights in livestock and livestock products. Women may not own livestock in the Western sense of 'property', but in the Maasai sense of 'property' they do. Within Maasai culture there is a strict separation between the notions of 'property' and 'control'. Several people (including women) may hold rights to the same animal simultaneously—and are therefore involved in decision making regarding its sale—but only one person has 'control' over the animal; thus rendering Maasai property rather diversified and complex.

This example of livestock management proves that it is not possible to draw a clear borderline between male and female domains. Rights and duties as well as influences have to be seen against the background of theoretical norms, and in the context of the actual situation and individual experience. In the present situation, however, the influence and the involvement of women in decision making concerning livestock transactions and land management is actually diminishing, creating a situation that corresponds ever more to the previously ideal gender relations.

The increasing importance of livestock cash transactions and the reduction of subsistence herds diminishes the partial control of women over livestock. The control that Maasai men have over their wives' labour is becoming indisputable, and they are getting a firmer grip on the distribution and allocation of livestock and livestock products (Talle, 1988: 250). Consequently the position of women within the pastoral production system is constrained by new principles of organization, i.e. land tenure and livestock marketing practices, which give precedence to male activity.

CONCLUSIONS

Gender relations mainly appear as static, and the dominant position of men is often taken as a status quo in the absence of analysis. Women are often presented in terms of their function within male dominated institutions. The lack of analysis of basic relations can often be explained by the fact that mainly male observers and researchers report. Gender relations and female tasks require more female investigation. At the same time, there is a need for self-presentation of socio-cultural perspectives, a need for confrontation with individual personalities and the multiplicity of social action. Outside observers always run the risk of including their own cultural backgrounds in their observations.

The presented method has proved to be effective with regard to the quantity and quality of information that can be collected. As soon as the women are used to sitting together and discussing community issues, they can be fully integrated into the process of evaluation, interpretation, and implementation.

In the present situation, monetization of the rural economy, increased need for cash income, and privatization of land lead to the diminishing influence of women in livestock management and therefore to limited rights within the society. This means a change in the actual situation of women, but not a change with regard to the ideal female status within Maasai culture. Even if women suffer from an obvious loss of freedom and self-determination, the present situation seems to be moving towards the ideal cultural norms.

Diminishing influence in economic processes and reduced access to family resources increase the need of women for other sources of cash income. This need, the increasing work load, as well as the apparent understanding of environmental issues and community affairs makes it necessary and worth while to involve women in the process of finding solutions to the problems. The aim should be to overcome women's separation from the public domain of production and decision-making.

NOTE

1. Translation:
   For the man, his wife is mainly important as a source of labour, looking after the household and livestock, and she is the guarantee for a high number of descendants.

She is treated according to the fulfilment of her duties. Moderate beating is not uncommon, brute force almost never occurs . . . The woman doesn't have any rights, neither with regard to the public domain, nor towards her husband.

REFERENCES

Ardener, E. (1977) Belief and the problem of women. In: Ardener, S. (ed.), *Perceiving Women*. London: Dent, pp. 1–18.

Arhem, K. (1985) The Maasai and the state: the impact of rural development policies on a pastoral people in Tanzania. *IWGIA Document 52*. Copenhagen, International Work Group for Indigenous Affairs.

Bay, E. and N.J. Hafkin (1976) *Women in Africa: Studies and Economic Change*. Stanford, CA, California University Press.

Burton, M. and L. Kirk (1979) Sex differences in Maasai cognition of personality and social identity. *American Anthropologist* **81**(4): 841–73.

Caplan, P. and J.M. Burja (eds), (1978) *Women United, Women Divided: cross-cultural perspectives on female solidarity*. London, Tavistock.

Collins, J. (1991) Women and the environment: Social reproduction and sustainable development. In: Gailin, R.S. and A. Ferguson Boukler (eds), *The Women and International Development Annual, 2*. Boulder, CO, Westview Press.

Evangelou, P. (1984) *Livestock Development in Kenya's Maasailand Pastoralists' Transition to a Market Economy*. Boulder, CO, Westview Press.

Fosbrooke, H.A. (1948) An administrative survey of the Maasai social system. *Tanganyika Notes and Records* **28**: 1–50.

Galaty, J. (1982a) Maasai pastoral ideology and change. In: *Contemporary nomadic and pastoral peoples: Africa and Latin America* (Studies in Third World Societies). Williamsburg, VA, College of Williamsburg, Dept of Anthropology.

Galaty, J. (1982b) Being 'Maasai', being 'people of cattle'—ethnic shifters in East Africa. *American Ethnologist* **9**: 1–20.

Hockings, Paul (ed.) (1975) *Principles of Visual Anthropology*. The Hague, Mouton.

Hollis, C.A. (1905/1970) *The Maasai: their Language and Folklore*. Westport, CT, Negro University Press.

Jacobs, A.H. (1973) Land and contemporary politics among the pastoral Maasai of East Africa. Paper presented at the American Anthropological Association Meetings *Symposium on Local Level Changes in Contemporary East Africa*, New Orleans.

Kipury, N. (1989) *Maasai Women in Transition: Class and Gender in the Transformation of Pastoral Society*. PhD dissertation, Temple University, USA.

Llewelyn-Davies, M. (1978) Two contexts of solidarity among pastoral Maasai women. In: Caplan, P. and J.M. Bujra (eds), *Women United, Women Divided: cross-cultural perspectives on female solidarity*. London, Tavistock, pp. 206–37.

Llewelyn-Davies, M. (1981) Women, warriors and patriarchs. In: Ortner, S.B. and H. Whitehead (eds), *Sexual Meanings: the cultural construction of gender and sexuality*. Cambridge, Cambridge University Press, pp. 331–58.

Merker, M. (1904/1910) *Die Maasai: ethnographische Monographie eines ostafrikanischen Semitenvolkes, 2. rev. edition*. Berlin, D. Reimer.

Mitzlaff, U. von (1988) *Maasai Frauen: Leben in einer patriarchalischen Gesellschaft—Feldforschung bei den Parakuyo, Tansania*. Munich, Trickster Verlag.

Talle, A. (1988) Women at a loss: changes in Maasai pastoralism and their effect on gender relations. *Stockholm Studies in Social Anthropology 19*. Stockholm, Department of Social Anthropology.

Part VI

GOVERNMENT POLICIES

To most of the people who live in the drylands, government policy is not something that is thought about. Many do not even know of its existence as a formal entity, yet they do experience its effects when it is applied to water development, credit facilities, infrastructure investments, livestock and agricultural produce marketing, the creation of protected conservation areas, land-tenure arrangements, administration structures and mechanisms and so on.

Government policy, and the effective implementation of that policy, has been identified as one of the most critical factors in creating an enabling environment that would permit local communities to decide and guide their own fate from the 'bottom up'. If government policy is to neglect dryland areas in favour of high-potential regions of the country, then the people living there not only have a harsh ecological environment to contend with but a harsh political one as well.

There are many policy decisions governments can take that would not overburden scarce treasury resources, and still facilitate good land management. Such policies revolve around the basic necessity of putting more power into the hands of local communities through giving them control of their land and resources and creating a sense of security.

17 Environmental Degradation and Public Policy in Latin America

JORGE E. UQUILLAS
The World Bank, Latin American Technical Department, Environmental Division

INTRODUCTION

Latin America is a continent with high geographical, biological, and cultural diversity. It has several mountain massifs like the Andes, the Planalto of Mato Grosso, and the Sierra Madre; dense rain forests such as the Amazon, the Choco, and the Peten; large savannas in South America, and important arid and semi-arid regions in Mexico, Brazil, and the Pacific Coast of Chile, Peru, and Ecuador. Its biodiversity is among the highest on earth; several 'hot spots' of biological diversity have been identified, particularly in the tropical regions of Central and South America: the Amazon region alone probably has over 5000 species of plants, 950 bird species, 200 mammal species, 300 reptile species and a similar number of amphibians, between 2000 and 3000 species of fish, and millions of insect species (Goulding, 1993). The socio-cultural diversity is illustrated by over 450 indigenous languages, belonging to a similar number of Amerindian cultures, and still spoken in the continent, in addition to the official European languages of Spanish, Portuguese, English, French, and Dutch. Latin American culture has benefitted from Amerindian, European, and African influences.

In contrast to its high environmental and socio-cultural diversity, Latin America presents some common problems of environmental deterioration and rather monolithic public policy approaches to solving the problems. This chapter thus deals with the declining quality of Latin America's physical and social surroundings and with governmental decisions that have contributed to it. It also analyses some public policy trends which would have beneficial consequences for the environment in the future.

ENVIRONMENTAL DEGRADATION

While Latin America still has large extensions of its area, such as parts of the high Andean plateaux and the Amazon region, in a relatively pristine state,

Social Aspects of Sustainable Dryland Management. Edited by Daniel Stiles

over the years there has been a process of degradation which is obvious. The degradation—defined as a loss of natural habitat—of Latin America has manifested itself in several forms, including deforestation, soil erosion, air and water contamination, reduced water catchment, drought and desertification, flooding, mineral depletion, and reduced carbon sequestration.

Along with the destruction of the physical environment, the socio-economic conditions of the continent have also deteriorated to the point that, in 1990, per capita consumption in most countries was less than in 1980 and more people were suffering absolute poverty in 1990 than in 1980 (O'Brien, 1991: 25). These conditions are the effect of processes of environmental degradation (Mahar, 1989; DeWalt *et al.*, 1993) and have undoubtedly contributed to it because they have forced people to over-exploit nature, regardless of the consequences (Mink, 1993; Baena Soarez, 1992).

Poverty is widespread in Latin America and affects urban as well as rural populations. However, analysis of the existing data shows that in some countries it affects more indigenous than non-indigenous peoples. For instance, indigenous peoples, who represent the highest proportion of rural population and significant proportions of urban population in Bolivia, Ecuador, Guatemala, Mexico, and Peru, are predominantly poor, due largely to an unequal distribution of resources in society, in terms of both income as well as of access to basic social services such as health, education, and housing. In Bolivia and Peru, where over half of the population is poor, between two thirds and three quarters of poor people are indigenous. In Guatemala, 66% of the population is poor, but 87% of the indigenous peoples are below the poverty line (Psacharopoulos and Patrinos, 1993).

Some aspects of environmental degradation have been estimated thanks to modern technology such as satellite images, radar, and aerial photography. Others are harder to estimate at the regional level. Nevertheless, *deforestation* is one of the main forms of environmental deterioration. In Latin America, which has 27% of the earth's forest cover (Browder, 1991: 45), between 1981 and 1985 there was an annual loss of tropical forest cover of 5.6 million ha, representing 0.6% of the remaining closed and open forest at that time, according to a study by the United Nations Food and Agriculture Organization (cited by Keipi, 1991: 39).

Tropical forests in Central America and the Caribbean have almost disappeared. But, in addition, the largest repository of tropical life on earth, the Amazon, is also being deforested at fast rates. By 1990, between 6% and 7% of its forests had already been cleared (Fearnside, 1989). In Brazil, where the largest part of the Amazon region lies, deforestation of the so-called 'Legal Amazon' has accelerated sharply since the mid-1970s; in 1988 the deforested area was about 600 000 km^2 and the total area cleared up to that year was over 5 million ha, largely due to the expansion of cattle ranching (Mahar, 1989). Likewise, deforestation of the Colombian Amazon has varied between 600 000 and 880 000 ha annually (Bunyard, 1989); the rate of deforestation of the Peruvian Amazon for the 1981–5 period has been estimated at 270 000 ha per year, much of it due to the expansion of coca plantations (Bedoya and Klein, 1992); and, at the present, the rate of deforestation of the Ecuadorian Amazon is over 100 000 ha per year.

Mexico provides an illustration of the deterioration of forests and other natural resources. In that country 60% of the rivers are seriously contaminated; the salinization of soils due to inappropriate irrigation models has reduced crop output by the equivalent of one million tonnes of grain a year, and wastewaters from urban run-off are threatening rich and complex coastal ecosystems. Likewise, about one-half of Mexico's land shows moderate to advanced erosion, and about one fifth has already been destroyed (Goldrich and Carruthers, 1992: 102–3).

Soil fertility is being seriously reduced in most of the continent. To start with, 81% of the soils are acidic and low in nutrients—as compared with 56% of Africa's and 38% of Asia's (Moran, 1987: 70). Adding to this *erosion* is affecting the mountain slopes of the Andes, the highland valleys of Central America and Mexico, and other fragile lands in the Amazon. Data on the amount of erosion are deficient, but one study in Honduras estimates that in the Upper Choluteca Watershed as much as 13 tonnes/ha/yr are lost to erosion (DeWalt *et al.*, 1993).

Land degradation and soil erosion as well as the severe impoverishment of the peasantry has been attributed to a model of 'disarticulated capitalism' which has led peasants to increase the exploitation of their resource base as a survival strategy (Faber, 1992a: 6–7). In more specific cases, settler impoverishment and environmental destruction in the lowlands of Bolivia are the products of 'unequal exchange relations' between the settlement area and the larger society (Painter, 1987: 165). This phenomenon has been also described as the 'simple reproduction squeeze', by which local populations, like the frontier people of Rondonia in the Brazilian Amazon, are forced to reduce household consumption and/or intensify commodity production to maintain a minimum subsistence level (Millikan, 1992).

Deforestation, like other forms of destruction of the physical surroundings, is attributed to many causes, among them population growth, poverty, unsecured land tenure, lack of technical and environmental awareness, lack of financing for small farmers, and debt (Tulchin, 1991; Mink, 1993). Most authors also stress that one of the main causes is public policy (Keipi, 1991; Browder, 1991; Southgate and Range, 1990). Even though worldwide the depletion of about 150 000 km^2 has been attributed to small-scale agriculture, Mahar (1989) points out that the small farmer is not the cause but the victim of deforestation. The real causes are likely to be poverty, unequal land distribution, low agricultural productivity, rapid population growth, and misguided public policies.

PUBLIC POLICY

Broad development policies

Public policy is here defined as the formally stated way to achieve the objectives set by a national, regional, or local government. It is useful to distinguish between policy which has indirect implications for the environment and that which is directly related to the environment.

Up to recent times there have not been environmental policies *per se* in Latin America, but instead policies that have had strong environmental implications. At the most general level, consideration needs to be given to the adoption of a given model of economic development or modernization, which has led countries to adopt some specific development objectives and strategies to achieve them. As differentiated models of development we have:

- Agro-exports, based on a plantation economy
- Import substitution industrialization
- Export-oriented mining, agriculture, and manufacturing.

Much of the land clearing in Central America is attributed to the rise of agro-exports based on a plantation economy. First, banana and coffee in the 1950s; then cotton and sugar in the 1960s; and beef cattle since the 1970s. The production of these commodities has generally been stimulated by international lending institutions and by development policies of modernization, economic diversification, and expansion of capitalist agriculture and industry. They have flourished at the expense of large areas of tropical forest in the lowlands. Central America's rain forest, one of the main reserves of biodiversity on earth, has been disappearing at the rate of 3500–4000 km^2 annually. Over one-fifth of land in the region is now under permanent pasture, representing more land than that dedicated to all other agricultural commodities (Faber, 1992b).

Of all the activities characteristic of a plantation economy, coffee processing has had the most serious effects on the degradation of water. For instance, in Costa Rica, the coffee beneficios have produced 66% of the water contaminants, including boron, chloride, and arsenic (Faber, 1992b: 22).

The import-substitution industrialization model of development, by favouring the manufacturing sector and the urban areas, also had some negative consequences for rural people and the countryside. The policies that promoted cheap food and labour induced a decapitalization of the rural sector, particularly of the peasantry. In an effort to maintain their incomes, some people had to produce more, either by expanding the land area or by over-exploitation of natural resources rather than improving the technology of production.

Currently, in Latin America, there are remnants of old models of development, but the trend is to promote exports in the context of an increasingly deregularized, more competitive international commerce. Its environmental effects cannot be fully anticipated, but there still exists a situation in which 'Developing countries of the hemisphere have confronted the need to use their natural resources in an indiscriminate manner in order to adapt to the demand of the developed countries for raw materials and primary goods, which have historically been their main exports' (Baena Soares, 1992).

The environmental costs of economic development have not been taken into account either by market economies or by those which adopted centralized planning. In fact, in the market-oriented economies of Latin America, 'nature' has been a factor of production and almost no attention has been paid to waste and pollution (Faber, 1992a: 5). The model of development followed by Latin American countries in the past, under the assumption that natural capital is infinite, has generated impoverishment, heavy migration from rural to urban areas, pressures on land use that have resulted in over-utilization and extractive exploitation of natural resources (Muñoz, 1992: 7).

Export production has particularly contributed to environmental loss by favouring internal processes of capital accumulation by a wealthy few based on the exploitation of natural resources. These people are usually not accountable for the social or environmental consequences of their actions; meanwhile, large numbers of people are denied or have only limited access to those resources and have no option but to over-exploit the few resources at hand.

Sector-specific policies

As strategies associated with one or more of the development models adopted and with strong environmental implications we have several more specific policies, among them:

- Agrarian reform and colonization
- Agricultural modernization.

These strategies involve policies that cannot always be desegregated. Usually they are interrelated and have a combined effect on the environment. Thus, policies which lead to government intervention in markets for agriculture and natural resource commodities, to inappropriate tenurial arrangements and inadequate investment in research and extension have contributed to tropical deforestation, soil erosion and the disturbance of coastal ecosystems in countries like Ecuador (Southgate and Whitaker, 1992).

In many instances, Latin American governments have adopted policies favouring deforestation. Southgate and Range (1990) point out four institutional incentives for deforestation: (1) open access, leading to waste and misuse of forest resources; (2) deforestation as a requisite for land tenure; (3) tenure insecurity; and (4) the demise of common property regimes.

In the 1960s, modernization efforts led to the adoption of agrarian reform and colonization policies. They promoted redistribution of land and labour by breaking up large estates, particularly the hacienda system. Some of the excess rural population was then routed to the growing urban centres and the rest to frontier areas, usually to rain forests in the lowland tropics that had previously been relatively undisturbed.

Instead of promoting the intensification of agriculture (higher yields through improved management of areas already under cultivation), governmental policies have often relied on frontier expansion or land extensions (Billsborrow and Ogendo, 1992) as a means of obtaining greater agricultural output.

In the majority of Amazonian countries, public policies also contributed directly to deforestation by promoting the clearing of tree cover as a condition of legal tenure and by a series of tax exemptions and credit incentives for pastures for cattle ranching and perennials and short cycle crops (Uquillas, 1984; Mahar, 1989; Binswanger, 1987). As a result, cattle ranching may be responsible for almost two thirds of the deforestation and environmental destruction that has occurred in the Amazonian regions of some countries (Mahar, 1989; Poveda, 1991).

High-input agriculture, generally associated with crop specialization, has led to changes in soil composition, reducing organic life, and has contributed to the chemical contamination of soils and waters, with severe consequences for human health. As previously mentioned, coffee processing has had serious effects on lands and waters, contributing to about two thirds of all water pollution in Costa Rica. In Honduras, the greatest threat to the Gulf of Fonseca is water pollution caused by the misuse of pesticides (DeWalt *et al.*, 1993).

Policies for the environment

Some specific environmental policies, adopted by governments in the last decade, have been oriented to control contamination or to protect natural resources, particularly forests. Nevertheless, there currently is a new set of public policies that have broad environmental implications and that is still in the process of being implemented. Some environmental laws have already been promulgated. They usually dictate the creation of public institutions in charge of overseeing the enforcement of such laws. To be viable, these institutions usually have required the creation of specific national funds for the environment. For instance, the government of Bolivia, in 1990, enacted a Law of the Environment, which led to the creation of the Secretariat of the Environment (SENMA) and of a National Environmental Fund (FONAMA), but the operative norms have not been developed.

Similarly, Chile, which previously had only a number of environmentally related provisions dispersed in its complex legal system, started in 1990 to design a policy based on (1) respect for nature; (2) environmental conservation; (3) energy conservation; (4) improvement in the quality of life; and (5) international collaboration to overcome worldwide environmental problems (Alvarado, 1992).

It is still too early to evaluate the impact of these new policies on the environment. In some cases, they have been considered unclear and ineffective. Usually, they need to have the corresponding regulations approved before they become operational, and that can take years, as the Bolivian experience demonstrates.

It has also been argued that environmental policies, to be successful, should address the political–economic roots of the ecological crisis rather than be limited to protect and restore the ecological conditions of capitalist production (Faber, 1992a).

PERSPECTIVES FOR THE FUTURE

The future of Latin America is going to be characterized by declining overall population growth rates but increasing numbers of inhabitants due to the high proportion of people of childbearing ages. Population growth is at present 1.7% per year and it is estimated that net increase in population will be at about 92 million people per year (Mink, 1993). With more population, particularly in the urban areas, there will be increasing demand for services, food, fuel, water, and other goods, placing additional stress on available natural resources.

Besides strict conservation measures for the remaining ‘hot spots’ of biological diversity in Latin America, strong public policies are needed to maintain the forested land. In this regard, it is important to keep searching for alternatives for forest protection and management, which include urban and rural residents, and particularly indigenous peoples in these efforts. The important role that the latter can play in the protection of forests has been acknowledged by the government of Colombia, which in the last few years has recognized indigenous peoples’ title to about 18 million ha of *resguardos*, not only in recognition of their land rights but also as a way of providing protection to the forests, in the form of thousands of forest guardians (Bunyard, 1989; Colombia, 1990).

Indigenous people of the Amazon have not only been the guardians of the forest. They have managed their forest and thus obtained from it the goods for their social and biological reproduction. For instance, the Lowland Quichua or Runa people of the Upper Napo Basin of Ecuador have traditionally managed the succession of the natural forest, enhancing food production while maintaining the forest apparently intact.

There is much that can be done to reduce degradation and stabilize natural systems in areas which are in the process of deforestation. One of them is the intensification of agroforestry systems. A case study of the Ecuadorian Amazon indicates that the use of improved agro-forestry technology is a real alternative for the management of secondary forests which would not have negative social or ecological implications. Improved agro-forestry favours a more productive use of labour, reducing also cash requirements during times of low prices for commodities such as coffee. In the short term, improved agro-forestry can reduce the current colonists' pressure on the resources of the natural forests by finding compensatory sources of income; in the medium term, such pressure will tend to disappear once the present wood inventory enters into a productive stage (Uquillas *et al.*, 1992).

In those areas of the continent which have been subject to continuous cultivation some of the increased food supply will come as a result of the intensification of production through the use of improved agricultural technology, particularly from high-yielding varieties of crops and animals (Mink, 1993). Important social and environmental goals will have to be achieved, however, by small-scale farmer production, oriented to household consumption and for internal markets, and widespread, labour-intensive environmental reconstruction (Goldrich and Carruthers, 1992).

Regarding land in tropical areas, it is argued that diversification of production rather than specialization around a comparative resource advantage has to be the central and guiding tenet of sustainable forest land use; this implies that if government subsidies are to be used, they should be directed to vertically diversified land uses instead of commodities (Browder, 1991).

Land reform and colonization will tend to be even less important as the privatization drive consolidates and available lands for new settlements become more scarce, with the exception, perhaps, of Brazil, which has extensive Amazon territories and where military circles are still concerned about 'threats' to national sovereignty.

Accompanying the urbanization process, or as a factor closely related to it, more housing and social services, transportation, and industries can be expected, and with them increasing waste and pollution. The trend towards adoption of environmental policies will consolidate. Colombia, Venezuela, Peru, Mexico, Ecuador, Bolivia, Costa Rica, and Chile have already approved environmental legislation or created national commissions or ministries to formulate and apply environmental protection codes. In fact, protection of the environment has become a development priority and a national security concern for many Latin American countries (Muñoz, 1992).

Greater awareness of environmental issues will also generate ever more active environmental social movements. These movements could range from those specifically concerned with the physical environment to those which promote greater empowerment of local populations and even those which advocate the peasants' appropriation of natural resources.

Inequality in the distribution of resources is at the heart of the problem of environmental degradation, including the wanton destruction of primary tropical forests in the Amazon countries. Greater equity, or the reduction of poverty, must be a primary objective of sustainable development before the issue of environmental quality can be fully addressed. Besides promoting poverty alleviation through increasing poor peoples' incomes it will be necessary to make targeted interventions to reduce the risks faced by the poor and secure their rights to natural resources and to improve their access both to resources such as land and credit as well as to essential services like education, public health, and family planning (Mink, 1993).

Misguided public policies have to change. In the case of Brazil, for instance, eliminating public incentives for livestock projects has been suggested, declaring a moratorium on the disbursement of fiscal incentives for projects which are over-exploiting timber from the natural forests, and modifying inappropriate land-granting policies (Mahar, 1989). Another author has stressed less road building in fragile Amazon areas (Fearnside, 1989); and this could be extended to other countries which want to reduce deforestation in tropical areas. A public policy that attempts to change the behaviour of powerful groups within a nation and that of lending and donor agencies has also been recommended (DeWalt *et al.*, 1993).

To conclude, environmental policies to date have focused on repairing damages previously caused. Now it is necessary to focus environmental problems on their origin in macroeconomic, commercial, and sectoral policies. Environmental policy must be an integral part of economic and social development policies. Its objective must be at least to prevent damage and to reduce the adverse effect of human activities and, in the best of cases, to actively promote a socio-economic policy that expands the base for sustainable development (Brundland, 1989).

REFERENCES

Alvarado, L. (1992) Environment and development: a view from Chile. In: Muñoz, H. (ed.), *Environment and Development in the Americas*. Boulder, CO, Lynne Rienner, pp. 19–24.

Baena Soarez, J. (1992) Foreword. In: Muñoz, H. (ed.), *Environment and Development in the Americas*. Boulder, CO, Lynne Rienner.

Bedoya, E. and L. Klein (1992) Coca expansion and environmental destruction in the Peruvian Amazon Basin: the case of the Upper Huallaga. Paper delivered at the *Seminar on Population and Deforestation in the Humid Tropics*. Campina, Brazil, 30 November–3 December.

Billsborrow, R. and H.W.O. Okoth Ogendo (1992) Population-driven change in land use in developing countries. *Ambio* **21**, No. 1 (February).

Binswanger, H. (1987) *Fiscal and legal incentives with environmental effects on the Brazilian Amazon*. Paper of the Agricultural and Rural Development Department, World Bank, Washington, DC.

Browder, J.O. (1991) Alternative rain forest uses. In: Tulchin, J. (ed.), *Economic Development and Environmental Protection in Latin America*. Boulder, CO, Lynne Rienner, pp. 45–54.

Brundtland, G.H. (1989) Herencia común, futuro común: medio ambiente y desarrollo. In: Cárdenas, M. (ed.), *Política Ambiental y Desarrollo*. Bogotá, FESCOL-INDERENA, pp. 19–26.

Bunyard, P. (1989) Guardians of the forest: indigenous policies in the Colombian Amazon. *The Ecologist* **19**, No. 16 (November–December).

Colombia, Republic of (1990) *The Policy of the National Government in Defense of the Right of Indigenous Peoples and the Ecological Conservation of the Amazon.* Bogota.

DeWalt, W., P. Vergne and M. Hardin (1993) Population, aquaculture and environmental deterioration: the Gulf of Fonseca, Honduras. Paper prepared for the Rene Dubos Centre *Forum on Population, Environment and Development*, New York Academy of Medicine, 22–23 September.

Faber, D. (1992a) The ecological crisis of Latin America: a theoretical introduction. *Latin American Perspectives* 72 **19**(1): 3–16.

Faber, D. (1992b) Imperialism, revolution and the ecological crisis of Central America. *Latin American Perspectives* 72 **19**(1); 17–44.

Fearnside, P.N. (1989) Deforestation in the Brazilian Amazon: the rates and causes of deforestation. *The Ecologist* **19**(6) (November–December).

Goldrich, D. and D.V. Carruthers (1992) Sustainable development in Mexico? The international politics of crisis and opportunity. *Latin American Perspectives* 72 **91**(1): 97–122.

Goodman, D. and M. Redclift (eds) (1991) *Environment and Development in Latin America.* Manchester, Manchester University Press.

Goulding, M. (1993) *Biodiversity and Amazon Floodplains: an ecological and environmental framework.* Unpublished manuscript.

Keipi, K. (1991) Reducing deforestation in Latin America: the role of the Inter-American Development Bank. In: Tulchin, J. (ed.), *Economic Development and Environmental Protection in Latin America.* Boulder, CO, Lynne Rienner, pp. 39–44.

Mahar, D. (1989) *Government Policies and Deforestation in Brazil's Amazon Region.* Washington, The World Bank.

Millikan, B. (1992) Tropical deforestation, and degradation, and society: lessons from Rondonia, Brazil. *Latin American Perspectives* 72 **19**(1): 45–72.

Mink, S. (1993) *Poverty, Population and the Environment.* Washington: World Bank Discussion Papers 189.

Moran, E. (1987) Monitoring fertility degradation of agricultural lands in the Lowland Tropics. In: Little, P.D. and M. Horowitz, with E. Nyerges (eds), *Lands at Risk in the Third World.* Boulder, CO, Westview Press, pp. 69–91.

Moscoso/Real/Haug/Enríquez/Acosta-Solíz (1991) *Ecología y Desarrollo.* Quito, FESO.

Muñoz, H. (1992) The environment in inter-American relations. In: Muñoz, H. (ed.), *Environment and Diplomacy in the Americas.* Boulder, CO, Lynne Rienner, pp. 1–12.

O'Brien, P.J. (1991) Diet and sustainable development in Latin America. In: Goodman, D. and M. Redclift (eds), *Environment and Development in Latin America.* Manchester, Manchester University Press, pp. 24–47.

Painter, M. (1987) Unequal exchange: the dynamics of settler impoverishment and environmental destruction in Lowland Bolivia. In: Little, P.D. and M. Horowitz, with E. Nyerges (eds), *Lands at Risk in the Third World.* Boulder, CO, Westview Press, pp. 164–91.

Painter, M. (forthcoming) Introduction: anthropological perspectives in environmental destruction. In: Painter, M. and W. Durham (eds), *The Social Causes of Environmental Destruction in Latin America.* Ann Arbor, MI, University of Michigan Press.

Poveda, J. (1991) *La Amazonia: conflicto de esperanzas y realidades.* Unpublished manuscript.

Psacharopoulos, G. and H.A. Patrinos (eds) (1993) *Indigenous People and Poverty in Latin America: An Empirical Analysis.* Washington, The World Bank, Latin America and the Caribbean Technical Department, Regional Studies Program, Report No. 30.

Southgate, D. and C.F. Runge (1990) *The institutional origins of deforestation in Latin America.* Staff Paper P90-S, Department of Agricultural and Applied Economics, University of Minnesota (January).

Southgate, D. and M. Whittaker (1992) Promoting resource degradation in Latin America: tropical deforestation, soil erosion, and coastal ecosystem disturbance in Ecuador. *Economic Development and Cultural Change* 40(4) (July).

Tulchin, J.S. (ed.) (1991) *Economic Development and Environmental Protection in Latin America*. Boulder, CO, Lynne Rienner.

Uquillas, J. (1984) Colonization and spontaneous settlement in the Ecuadorian Amazon. In: Schmink, M. and C. Wood (eds) *Frontier Expansion in Amazonia*. Gainesville, FLA, University of Florida Press.

Uquillas, J., A. Ramirez and C. Sere (1992) Are modern agroforestry practices economically viable? A case study in the Ecuadorian Amazon. In: Sullivan, G., S. Huke and J. Fox (eds), *Financial and Economic Analysis of Agroforestry Systems*. NFTA–FSP–East West Centre–USAID, pp. 273–92.

18 Government Policies to Promote Good Dryland Management

DANIEL STILES
Nairobi, Kenya

INTRODUCTION

Six government representatives were invited to the 'Listening to the People' workshop with a request to present their governments' policies dealing with the following questions:

- What are your government's policies regarding rural communities in terms of land tenure, access to resources on common property, decentralization of political decision making, and provision of extension services and credit schemes to assist agricultural development?
- How well is this government policy implemented? What are the constraints or obstacles to implementation and how might they be overcome?
- How much government involvement is required in your country in the conception, formulation and implementation of rural development and environmental conservation projects?

The papers presented by government representatives varied considerably in their response to these questions. A summary of each is presented here.

INDIA

The paper was presented by Mr T.K.A. Nair, Additional Secretary in the Department of Wastelands Development, Ministry of Rural Development.

Land reform and ownership rights

The major objectives of land reform are to eliminate exploitation in land relations and make ownership more egalitarian, abolish intermediaries, enlarge the landholdings of the rural poor, and to increase agricultural productivity. The strategy includes a ceiling on ownership of agricultural holdings, consolidation of holdings, distribution of government wastelands, conferring ownership

Social Aspects of Sustainable Dryland Management. Edited by Daniel Stiles

rights to landless people, the protection of land of Scheduled Castes and Tribes, improving land access by women and safeguarding common property resources.

Statistics were presented showing the results thus far of implementation of the policies. Large landholdings (>10 ha) have come down while the number of smallholdings (<2 ha) have gone up. With the abolition of intermediary tenures some 20 million cultivators now deal directly with the government. Up to June 1992 some 4.8 million beneficiaries have received land. Computerization of land records has aided consolidation and distribution efforts. Land reform is being integrated with other national rural development programmes to improve the living standards of the rural poor. The Eighth Five Year Plan (1992–7) aims to fulfil the objectives of land reform policy.

Common property access

Common Property Resources (CPRs) in India consist of community pastures and forests, wastelands, common dumping and threshing grounds, watershed drainages, village ponds and rivers and streams (including their banks). CPRs are a particularly significant resource in the drylands, providing a wide variety of food, fodder, fuel, construction materials, trade items and craft materials, water, and so on.

The collapse of traditional management systems and the increased privatization of land is leading to the deterioration of CPRs. No specific policy towards improving CPR management was presented.

Decentralization of political decision making

The Constitution Bill (72nd Amendment) passed by parliament in 1992 had the main objective of ensuring people's participation in the process of decision making. The Amendment calls for Gram Sabhas (councils) in each village and Panchayats at village, intermediate and district levels, with reservations for minorities and women. The Panchayats would be elected and hold office for five years. The Panchayat Raj system varies from state to state, but the Amendment provides for empowering them to levy taxes/duties and for the setting up of a Finance Commission every five years to review the financial position of the Panchayats and make recommendations to the state governments.

Extension services

In order to improve the transfer of new agricultural technologies from research to farmers, and to gather feedback from farmers with which to orient future research, the Indian government has introduced the 'Training and Visit' system of extension. This system is operating in seventeen of India's twenty-one states. Initial evaluation of the system reveals that there have been positive institutional and economic changes.

Rural credit schemes

The central and state governments are working to provide credits to poor rural people through the Integrated Rural Development Programme and a pilot Family Credit Plan to save them from exploitation by moneylenders. Commercial and regional rural banks are participating to advance small loans to poor people to enable them to invest in enterprises and tide them over emergencies. The results of the scheme thus far are quite encouraging, though only a small proportion of the rural population is reached. The pilot phase will be expanded in future.

Policy for degraded land development

Of India's 329 million ha of land, about 130 million ha have been classified as wastelands or degraded lands producing below their original potential. The National Wastelands Development Board (NWDB) was established in 1985 to implement the Wastelands Development Programme, which aimed to rehabilitate degraded land and implement policies to promote good land management. In 1992 the Department of Wastelands Development was created and positioned in the Ministry of Rural Development. The NWDB was placed under this new department, and a National Afforestation and Eco-Development Board was set up to look after forest lands.

The mandate of the Department of Wastelands Development is the development of non-forestation wastelands for the sustainable production of biomass for energy and fodder. The department will work closely with local communities and NGOs to develop an integrated and participatory approach to wastelands development.

The changing role of government

The government of India has learned from experience that people's active participation in development efforts is necessary for success. The Eighth Five Year Plan proposes to make development a people's movement. The focus will be on creating or supporting existing people's institutions that are accountable to the community for improving development delivery systems using the vast potential of the voluntary sector.

The role of government will be to facilitate people's involvement in developmental activities by creating an appropriate institutional infrastructure. Voluntary and community organizations will be strengthened and supported, particularly the village-level Panchayat Raj institutions.

THAILAND

The paper was presented by Mr Chartree Chueyprasit, the Director of the Natural Resources and Environmental Management Division of the Office of

Environmental Policy and Planning, Ministry of Science, Technology and Environment.

An average annual economic growth rate exceeding 10% since 1987 has been achieved at the expense of natural resources and the environment. Urban industrial expansion has resulted in high rural–urban migration, but population growth in rural areas has nevertheless resulted in agricultural expansion into forest lands. Some 8–9 million ha of forest land have been encroached on by landless farmers, who abandon plots after two to three years and move on. More than 10 million people are living in forest reserves. The Thai government recognizes that there are serious problems of deforestation, soil erosion, and degradation of watersheds. About half of the country's agricultural land is degraded to some extent.

Land tenure

Land management is affected by insecurity of land ownership, which deprives farmers of both access to credit and the incentive to improve their land. Currently about 600 000 farmers are renting 3.07 million ha of farm land from landlords and speculators. Public land is very unequally distributed, and 30% of private lands have no legal documents. Land tenure and land allocation are a serious national problem.

Land tenure reform is being implemented through the Land Reform Act 1975 and other supporting measures. Land reform aims particularly at distributing farm land to small and landless farmers. Development activities accompany land-allocation measures. Public land allocation and development is funded under two programmes, the first under the national budget and the second in part or totally with loans or foreign assistance. Private land is also purchased and allocated.

Land reform has not gone very well, however, and in almost twenty years less than 40% of allotted land has been allocated. Constraints that have been identified are:

- Lack of unified policy and procedures. At present, more than fourteen agencies are carrying out land-allocation and settlement projects.
- Land policy has not been harmonized with forest policy, making the forest-encroachment problem harder to solve.
- There is confusion about which areas should be handled by which development programmes.
- The people affected do not always co-operate.

Recommendations to improve the land reform process included better organization of the agencies involved, the acquisition of additional land for distribution, more structured procedures for allocation, the issuing of full standard land titles rather than the multitude of document types currently given and improved financing measures.

Other questions

Questions of popular participation, decentralization of decision making, provision of credits, etc. were not addressed in the paper. It was stated that the Seventh National Social and Economic Development Plan (1992–6) aims to promote more equitable income distribution and rural development as well as enhance natural resource management.

SYRIA

The paper was presented by Dr Adnan Shuman, Director General of International Consultants United, Aleppo.

Land tenure policies

Syrian policies regarding land tenure and agricultural services have changed over time. In 1958 and 1963 programmes of land and agrarian reform began that resulted in a more equitable distribution of land. The feudal system of large absentee landholders was broken up and holders with more than 100 ha of land were reduced from 49% to 18% of the total, with corresponding increases of holders with smaller land areas. In 1965, Peasants' Unions were set up at village and higher order levels to represent the communal concerns of rural villagers. In 1970 policies were liberalized to promote participation of rural organizations in the formulation and implementation of rural development activities. Rural development became more important and attention focused on modernizing agriculture, improving extension services, and giving more attention to education, health and other social services.

Nothing specific about current land tenure arrangements was presented, but recommendations for improvement were made:

- Land tenure laws and regulations should be reviewed in order to secure and stabilize land tenure relationships.
- Fragmented holdings should be unified.
- Agrarian reform and land title registration should be completed quickly.
- Some responsibilities for land reform should be transferred from the Ministry of Agriculture and Agrarian Reform (MAAR) to the Peasants' Union.

Access to CPRs

The situation in Syria concerning access to common property resources appears to be complex. On the one hand, the paper stated that access to CPRs was governed by farmers' societies with assistance from the Peasants' Unions with a limited government role. On the other, there are national laws that restrict access with the stated objective of protecting the environment and preventing degradation. The scale of these CPR restrictions was not stated.

Another type of CPR is the pasture lands used by the nomadic Bedouins. The Badia (desert) Directorate within the MAAR was set up to deal with the nomadic groups. The policy is to sedentarize them in order to provide social services and security and reduce land degradation. In addition, in 1988 the government decided to eliminate the traditional usufruct rights of pastoralists to prevent them from cultivating steppe areas, which causes degradation.

The Directorate of Forestry in the MAAR plays a protective role for the remaining natural forest, preventing local utilization of resources. It also promotes government-sponsored afforestation programmes.

Decentralization of decision making

The Peasants' Unions described above were supposed to devolve decision making on rural development and natural resource management issues to local communities. The central Ministry of Local Government, however, still appoints governors, mayors, and all other civil servants. These officers are all directed by and responsible to the central government, not to the local people. This lack of democracy is explained by the government as due to the state of war of the country with Israel. It seems clear that the central government overrides any local management of CPRs and that local resources are denied access under the guise of environmental protection.

Extension services

Although a directorate for extension services was established within the MAAR in 1978, extension services are still weak. Agricultural extension workers co-operate with rural organizations to introduce new technologies and to improve the ability of farmers to solve their own problems through training and education. There is now a network of 754 extension service units in Syria.

Agricultural credits

The Agricultural Co-operative Bank (ACB) is the only official government lending agency in Syria. Policy is geared towards providing short-, medium- and long-term loans to agricultural co-operatives to meet their members' needs for inputs at subsidized interest rates. The ACB also acts as an instrument to channel government and international donor assistance for specific projects. Although the ACB lent out more than US$ 300 million to the public, co-operative and private sectors between 1986 and 1991, it still could not satisfy demand. Commercial and industrial banks are also approached for credit, particularly for farm machinery.

Constraints to policy implementation

- Land tenure—Landholdings are scattered, with each landholder having an average of 4.5 different plots. Some are too small to farm economically and they are rented or sold to large landholders, defeating the purpose of land reform. The government needs to intervene to rationalize landholdings.
- Land degradation—Population pressures and a dry climate are working to degrade land, lowering agricultural productivity and the availability of biomass resources. The government should involve the people more in the formulation and implementation of rural development and natural resource conservation projects.

Government involvement

Presently, the central government initiates and plans development on the grounds that it represents the interests of all groups. The Peasants' Union is consulted later, though the MAAR makes all final decisions. Dr Shuman interviewed MAAR senior officials on their views on rural development, and he noted that they were adopting a new way of thinking regarding community participation and collaboration. The paper recommended that real participation of rural people be promoted that would take into account their legitimate rights to access to their natural resources. This should be done in partnership with government through democratic processes. How much these recommendations were reflections of official government policy or Dr Shuman's personal views was not clear.

MALI

The paper was presented by Mr Yacouba Doumbia of the Ministry of Environment.

Land tenure

In pre-colonial times farm land was managed by family lineages under the overall administration of local chiefs or kings. CPR grazing lands were also managed by lineage heads following traditional customary law. In 1904 the French colonial government declared all 'vacant' land the property of the state. At independence in 1960 the Malian government assumed control of all land not under cultivation. Village lands were left under customary management.

In 1973–5 the government started up Rural Development Operations (RDO) in homogenous ecologically defined territories. There are now fifteen RDOs, in which development policies are implemented through extension, marketing, and credit organizations that enjoy financial and management autonomy. The paper did not state whether the RDOs operated on state and/or village lands.

The government came to the conclusion that the RDOs did not involve enough people's participation, and that uncontrolled agricultural and livestock development, in conjunction with chronic droughts, were resulting in land degradation. The government decided to begin a process of decentralization and democratization of Malian society.

Decentralization

In 1993, Law 93-008/AN-RN redefined the basic principles of administrative and territorial organization. The state began the process of transferring decision making and management powers to local collective structures called *Communes*. A Commune is defined as a grassroots administrative collective grouping persons having mutual interests and group solidarity. A Commune could be made up of urban neighbourhoods, village groups, or pastoral 'fractions'. Each Commune elects its own council.

A grouping of Communes, both urban and rural, forms a *Circle* (district). Each Circle forms a decision-making body made up of elected members of the Commune councils. A *Region* is the next highest-order geographic and demographic entity, made up of Circles. A regional assembly is composed of the elected representatives of the Circle councils.

The new land policy involves a progressive disengagement of the state from centralized production and marketing activities, and the promotion of private and collective development activities. The various local and regional elected councils will now increase the responsibilities and participation of local communities in development and natural resource management activities.

Constraints

The success of the new policy depends on revising the land code and clarifying land ownership and usufruct rights. Currently, there are many disputes about which laws (traditional, colonial period, or recent) have priority application and who should have management control of various land parcels. The new collective groupings also need to be strengthened and their juridicial and financial roles clarified.

BOTSWANA

Mr Seeiso Liphuko, Executive Secretary of the Natural Resources Board, presented the paper.

The principal land use and economic system in Botswana is livestock production. It is also the main cause of land degradation. The following discussion therefore centres on grazing lands.

Land tenure

Botswana has three basic systems of land tenure: state land, freehold private land and tribal communal land. Pasture lands are a CPR, except in the Tribal Grazing Land Programme farms and on freehold farms. About two thirds of national grazing lands are communal areas, and it is these areas that are undergoing the most degradation. To facilitate investment and land improvement, the government has introduced a system of Common-Law Lease on communal Tribal Land. An individual or group of users can make inputs (e.g. water development) and obtain a leasehold title. The title can be used to raise resources for further investments, or it can be sold. The sale applies only to the improvements, it does not confer ownership on the land.

Common property resources

Water is the main CPR. Even when boreholes or dam reservoirs are privately owned, water must be provided to those in need for personal or livestock use. Private water sources are normally used for a minimal payment. The government carries out water development on communal lands, which is free to users. There is a need for better land and water management to prevent degradation related to water points located in ecologically vulnerable areas. Uncontrolled water development is resulting in the growth and spread of cattle into areas where overstocking is leading to overgrazing.

Extension services

These mainly involve extension agents who aim to improve livestock quality through bull-subsidy schemes, artificial insemination, and veterinary care. Other subsidies and low taxes encourage livestock production.

Public participation

The basic grassroots forum is the traditional village council, or *kgotla*, in which any adult male or female may speak. This is a democratic, transparent body which is well understood by local communities. A local chief or headman chairs it and it forms part of the official tribal institutional framework.

Non-traditional public participation bodies are Village Development Committees, District Conservation Committees, Farmers Committees, Parents–Teachers Associations and 4B Clubs (Ministry of Agriculture) which are used for information exchange and to obtain community input into district and national policy formulation. The role of women is significant in these bodies and they are often in leadership positions. The Village Development Committees make contributions to the formulation of the National Development Plan.

The government encourages the formation and participation of NGOs in the development and conservation process and co-operation between government functionaries and non-governmental organizations. Local committees and NGOs contributed to the five-year process of preparing the National Conservation Strategy.

Policy constraints

Most of the constraints to achieving sustainable resource management revolve around traditional values related to livestock. Livestock, mainly cattle, have cultural as well as economic value. Large herds result in social rewards as well as economic ones for the owner. The large herds, however, can result in environmental damage.

The government has the power under the Agricultural Resources Act to confiscate and sell cattle deemed in excess of the land's carrying capacity, and reimburse the owner their value. The law has never been implemented, however, as it would cause extreme alienation among the people, and the owner would only buy more cattle with the sale proceeds.

Communal land and water resources which have cheap access is also seen as a constraint. As long as these resources cost little to users, there is an incentive for over-exploitation.

These constraints can be overcome by government policy incentives and disincentives. For example, over-stocking can be reduced by having the Botswana Meat Commission offer bonuses in degraded areas based on the number of animals sold. Water prices should be raised based on the number of animal users. Projects should be formulated and implemented that create awareness about the long-term consequences of over-stocking, but which also create financial surpluses through private sector involvement that can be used for further development and conservation activities. The government would act as a facilitator in this process, with local participation taking the lead.

MEXICO

Dr Manuel Anaya-Garduño, Research Director of the Soils Department, Graduate College, Montecillo presented the paper. The paper did not address any of the questions posed, but rather focused on the physical and technological aspects of the land degradation problem in Mexico.

CONCLUSIONS

Most governments seem to be aware of the growing concern for community participation and decision-making power related to natural resource management. Many governments are taking concrete steps to decentralize and devolve power to regional and district-level bodies, though others are still reluctant to do this. Land tenure laws and practices still form a major area that is in need of attention. Although the papers said little about it, there is also a need to improve communications and understanding between government officials and extension agents and local people.

It is at least encouraging to note that the social issues raised in the questions put to the invited governments are now on the agenda and are being seriously discussed and, in many instances, addressed. The ball is rolling and over the next several years it seems probable that more power will be transferred from central governments into the hands of the communities most directly concerned with good land and resource management. Local decision making and management can only work for sustainable development, however, if those people are well informed and competent for the task.

Part VII

CONCLUSIONS

Workshop Recommendations

Desertification and environmental degradation are complex processes brought about by a varied mix of interactions between political, social, economic as well as natural factors at global, regional, international, national, and local levels. The outcome of these processes are highly varied and location-specific. It is, therefore, impossible to devise general prescriptions for these highly complex problems.

Mounting evidence suggests that the introduction of development intervention itself has, at times, been a contributing factor in the various processes leading to desertification. Some of the major causes for these are: increasing pressures on land at the local level, the introduction of inappropriate development interventions such as top-down planning and implementation, inappropriate technology (for example, large dams) and an overly optimistic drive of planners to enforce sedentarization of pastoralists, the impacts of insecurity in land tenure, the local effects of structural adjustment programmes, and armed conflicts. Peace is an essential prerequisite for development, not only for the stability of government but also to avoid wasteful expenditure on war which puts additional pressure on natural resources. Population pressure, large-scale migration, and increasingly frequent droughts are also important factors. The widely acknowledged link between poverty and environmental degradation suggests that a fundamental element in good environmental management is the alleviation of poverty.

We affirm that any meaningful strategy to address these problems will have to result in fundamental changes in power relations between the various actors at international, national, and local levels. Especially the most directly affected groups of people in drought-prone areas themselves will have to be given channels to express themselves and negotiate with representatives of other groups and governments whose interests and activities are having impacts on their livelihoods in the areas affected by desertification in order to compromise between their and these other actors' interests at international and national levels. Nothing short of radical changes will lead to effective solutions and lasting results.

A massive redirection of the development effort away from a top-down directive to a democratic approach responsive to the specificity of local situations and needs is required. This should be an integrated approach that will balance conservation with local survival and social development needs,

Social Aspects of Sustainable Dryland Management. Edited by Daniel Stiles

prioritizing local needs for sustainable livelihoods over national needs and interests of various kinds. It must promote the self-empowerment of local peoples on a large scale, so that at all levels they have the right to participate in the decision-making process and express themselves freely and without fear. NGOs that work at grassroots level and are accountable to the people are appropriate change agents that can be used to make people aware of their rights and obligations.

From the background of our research and experience we acknowledge that many of the traditional practices, organizations, production-, consumption- and marketing mechanisms, indigenous knowledge and strategies of local peoples for coping with environmental change have been highly sophisticated and adapted to local situations. However, many of these have been lost and broken down due to mounting pressures outside local people's control. We advocate the preservation and—wherever possible—revival of these survival systems as the basis for locally adapted solutions to specific local food security problems and processes leading to desertification in general.

For effective strategies to combat desertification we provide the following recommendations.

DONOR AGENCIES AND INTERNATIONAL BODIES

(1) Donors and international bodies should encourage and support efforts towards more effective interaction among all the actors in the development process.
(2) There should be a change in emphasis from projects to programmes that are more popularly based, flexible and responsive.
(3) International bodies should review their structure and procedures in the light of the principles of this document.

GOVERNMENT POLICIES

(1) A prerequisite for any kind of sustainable development policy is a popularly based development strategy.
(2) Trade, price, credit, social, and economic policies should be evolved with participation by local communities (including indigenous peoples) and other sectors of society.
(3) Governments should be encouraged to fully decentralize policy formulation and decision making and streamline bureaucracy.
(4) Governments' environment policies should be formulated in co-ordination with all sectors and institutions concerned.
(5) Local communities should be involved in deciding what national and international research and development programmes should be undertaken in their areas, and in the environmental impact-assessment procedures.

(6) Governments should legitimize the existence of community groups by guaranteeing them freedom and autonomy in the management of their natural resources and management programmes, and co-ordinate this process of legitimization with neighbouring countries to account for transboundary pastoral movements.
(7) Governments should renew their efforts to eradicate illiteracy in rural areas, and ensure that environmental education is automatically included in the curriculum.

LAND TENURE

(1) The priority reform in regions where customary communal tenure systems are still in existence is for the state to recognize, through legislation, customary land rights and land tenure relationships.
(2) Land reforms aimed at providing equitable access to land by those actually working it and providing them with secure, clear rights (and obligations) associated with their land tenure are essential, although not sufficient, for improving natural resource management.
(3) The kinds of land reforms most appropriate for specific countries and situations have to be worked out locally, in consultation with all actors ensuring that women, mobile pastoral groups, and other groups at risk are included in the process.
(4) The international donor community has a role to play in exerting influence, by establishing criteria of conditionality where national governments are unwilling to accord local communities appropriate land tenure that promotes good land management.
(5) Land clearing should not be a precondition of land tenure rights.
(6) Land tenure policies should take into account the rights of access to grazing and water customarily exercised by livestock producers, which should be documented and officially recorded.

PARTICIPATION

(1) Realizing that there are diverse interests at the local level, marginalized groups should be supported by government and NGOs to organize themselves to represent their interests.
(2) Local communities should be involved in information collection and should identify the problems that are most in need of research with a priority oriented towards finding solutions to natural resource management and food production problems.
(3) Culturally appropriate participatory methods should be used at all stages of the development process to ensure that interaction between local people and development agencies is ongoing.

(4) Local communities should be kept informed by their government about all projects, policies, and programmes that will affect them.
(5) Although participatory methods are recommended, care has to be taken that a full commitment is made to genuine participation by all sectors of the community.
(6) Technological solutions derived from participatory research should be in line with the socio-economic, ecological, and political contexts and management systems of the local communities concerned. Any such technologies must be economically sustainable and environmentally non-threatening.

INDIGENOUS KNOWLEDGE SYSTEMS

(1) A prerequisite to planning any intervention in a local community area is to understand and recognize indigenous knowledge and management institutions and structures. Opportunities should be provided for development workers in collaboration with local communities to learn methodologies for identifying and documenting indigenous knowledge.
(2) An analysis and understanding must also be achieved of indigenous conceptions of existing systems of production, consumption, and marketing. These conceptions often include socio-cultural factors ignored or not noted by outside agents of the development process.
(3) A community's perceptions and beliefs relating to the multiple processes that lead to environmental degradation need to be understood in order to plan effectively to alter any negative factors. Emerging out of indigenous and experts' knowledge regarding the problems and solutions, proper management techniques should be developed which protect biodiversity in different ecological zones.
(4) Indigenous knowledge studies need to be carried out by competent local people.
(5) International agencies, governments, NGOs, and local communities should carry out inventories of economic and natural resources based on indigenous knowledge (IK) of males and females of all ages and on scientific research. This information should be compiled in a database, to be accessible and widely disseminated at all levels. Efforts should be made to make this information available in local languages as much as possible.
(6) The resultant indigenous knowledge information should be incorporated into any formulation and implementation of programmes and project activities.
(7) Incentives should be provided for farmers and pastoralists to conduct informal experiments: the formation of informal organizations of farmers and pastoralists should also be encouraged.

BIODIVERSITY, CONSERVATION, AND SUSTAINABLE DEVELOPMENT

(1) Special efforts need to be made to protect and enhance genetic diversity in wild plants and animals, livestock, and domestic crops. It should be recognized that indigenous peoples' subsistence systems are often repositories of unique genetic variability found in domestic plants and animals.
(2) It is recommended that technological diversity be promoted in order to deal with the biodiversity found in cultivars and plants found in the natural vegetation, and also local variability in soils and soil moisture conditions.
(3) Development agents aiming to control land degradation in drylands should accept technological diversity as a strategic aim, recognize and support indigenous adaptive technological development, as well as searching for new or improved technologies.
(4) The international community and governments should support efforts by local communities to develop environmentally sustainable marketing systems of renewable natural products from threatened dryland ecosystems to counter the prevalent exploitative activities that tend to degrade natural ecosystems.
(5) Intellectual property rights relating to indigenous knowledge and technologies should be protected through the application of the relevant sections of the Biodiversity Convention.

GENDER ISSUES

Equitable access to resources is a key element in the promotion of sustainable development. Despite this, the situation of many women in dryland areas is deteriorating, as their rights and access to productive resources is undermined and demands on their labour increase. Women have demonstrated considerable resourcefulness in adapting to the largely detrimental impact of these social, economic and environmental changes that are taking place in dryland areas.

(1) Any strategy to improve and safeguard the local environment should build on the knowledge and resourcefulness of local women, and address their specific needs.
(2) Governments should legislate to ensure that women have equal ownership rights when common property resources are privatized.
(3) It should also be recognized that development thinking and practice are usually gender-biased and discriminate against women. Women should be enabled to participate fully throughout the planning process, and to set their own priorities. In addition, women should be recruited into extension and project management at all levels.

(4) Gender sensitive training should be provided to male and female staff at all levels.
(5) Mechanisms for promoting women's equal participation in the decision-making process at all levels should be supported by strengthening existing women's groups and organizations, or where the need is expressed to assist with their formation.
(6) Where women themselves identify the need, time spent on domestic labour should be reduced.
(7) Women's productivity and independent income-generating capacity should be enhanced by supporting both traditional and alternative income-generating activities.
(8) Efforts should be made to ensure that women retain control of their earnings by providing savings facilities specifically for women.

ENVIRONMENTAL REFUGEES

(1) States are urged to pay particular attention to the increasing stream of internal and cross-boundary displaced people for environmental reasons.
(2) States are also urged to focus on mitigating the driving forces behind such displacements and to recognize coerced cross-border movements caused by environmental disruption or long-term degradation as a severe human problem.
(3) Regional arrangements should be developed for protection of and assistance to such displaced people.
(4) Affected countries should recognize that (a) risk of primary production failure is a distinctive part of dryland environments; and that (b) drought and land degradation, although sub-national in extent (in some countries), is a matter of international concern, justify an internal response, in particular to minimize population movements ('environmental refugees') between countries.
(5) Countries in a position to provide assistance should recognize that the mitigation of unpredictable food crises in dryland areas of affected countries is a legitimate form of desertification control, as it (a) protects household capital for land-improving investments and (b) minimizes disruptive population movements.
(6) Affected countries' governments should aim to support and enhance the adaptive capabilities of smallholder production systems in drylands (e.g. crop diversification or rotation, indigenous trees and irrigation technologies in farming systems), and to minimize the disruptive effects of new technologies and systems on the capabilities.

COMMUNICATIONS AND INTERACTION

(1) New social relationships and channels of communication need to be developed to facilitate not only participation by local communities in the

development process but also interaction between local communities, NGOs, governments, and donor agencies.

(2) All actors involved in dryland and natural resource management should be equal partners, each with something to contribute and something to gain from successful interventions.

RECOMMENDATIONS FOR FOLLOW-UP ACTION

(1) UNEP should consider establishing a small working group made up of social scientists and NGO representatives to advise on follow-up action that will be oriented towards implementing recommendations made at this workshop.

(2) UNEP should consider initiating and co-ordinating a series of pilot projects and activities that would incorporate and demonstrate implementation of the various recommendations. These projects and activities should involve as much as possible participatory methods of problem identification, planning, and implementation and aim to improve current methods of communication and interaction between local communities, NGOs, national governments, and donor/technical assistance agencies.

(3) If the above recommended follow-up steps are taken, UNEP should seek government, NGO, and other international donor partners with whom to carry out the actions in partnership.

(4) It should be recognized that the project formulation and planning phase is as important as the implementation of the activities. To carry out fully many of the recommendations contained in the sections above a great deal of time may have to be spent in participatory approaches and the gathering of community-based information, including indigenous knowledge.

(5) Given a recognition that many government and donor agency personnel are not familiar with the socio-cultural factors affecting natural resource management and dryland degradation, it is recommended that appropriate NGOs and social scientists be encouraged to develop sensitization and training programmes oriented towards improving socio-cultural understanding among personnel involved in the development process. Governments and donor agencies are encouraged to support these efforts and eventually make them part of standard requirements and procedures within their respective organizations.

(6) More attention must be devoted to research on socio-economic aspects of dryland management: researchers and knowledgeable NGOs should make a concerted effort to demystify myths concerning pastoral and forager societies and economies through publications, workshops and interactions with development agents.

(7) UNEP should encourage the development and dissemination of new and existing methodologies of participation and interaction.

Index

UNIVERSITY OF NOTTINGHAM
10 0428344 9
WITHDRAWN
FROM THE LIBRARY

UNIVERSITY LIBRARY
- 7 JUN 2006
SEM HALL 22
UNIVERSITY LIBRARY
- 7 JUN 2006
SEM HALL 32
UNIVERSITY LIBRARY
- 7 JUN 2006
SEM HALL 32
23 MAR 2009